Hands-On Science

Forensics

Brian Pressley

1 2 3 4 5 6 7 8 9 10

ISBN 978-0-8251-6515-3

J. Weston Walch, Publisher

40 Walch Drive • Portland, ME 04103

www.walch.com

Printed in the United States of America

Contents

To the Teacher

The investigation of a crime scene is one of the most complicated endeavors a scientist can undertake. The number of disciplines that an investigator must be familiar with is huge. The work of collecting evidence requires a methodical approach and might require a strong background in chemistry, physics, biology, geology, and any number of areas where these fields overlap.

As a teacher, you might wonder how a book on forensics is useful to you in your classroom. You may choose to use some or all of these activities. You will find that each one is a stand-alone activity, but the two summary lessons at the end are best done after many of the earlier activities have been completed. Most of the activities are meant to be finished in a single class period, although some can be made as detail-oriented as possible and can be spread across a few classes. The material covered would fit well into a general science class, and various activities would be fitting for a biology, a chemistry, a physics, or an earth science class.

There are television shows that feature forensics so intensely that the science seems like a character. Many students are captivated by the mystery that must be solved. Most shows are careful to have experts on the set to be sure that the science is right, and many of your students will be familiar with the basics. The goal of this book is to advance the scientific-thinking skills of your students by turning students into good scientists. The activities employ a number of process and inquiry skills, much like those used by a good scientist. Good scientists will:

- accept well-tested findings
- utilize time-tested procedures to produce accurate results
- see the things in front of them and not what they want to see
- change their minds when faced with new experimental evidence
- be skeptical of information but open-minded to the possibility that it is correct

Much of science is a combination of problem solving, the scientific method, and just plain hard work. Forensics might be the ultimate combination of all three. Correlation charts for the activities in this book connect each activity to the National Science Education Standards for grades 5–8 and 9–12. The correlations are listed again within each activity. You will also find a General Rubric to help you with assessment. Properly used, this book can be fun for students while still helping them strive to reach the high standards that you expect from them in your classroom.

National Science Education Standards Correlations

The National Science Education Standards are not presented in standard outline form, so referencing a particular item is difficult without presenting the entire item. A summary of correlations follows. The correlations are separated into grades 5–8 and grades 9–12. The two charts allow you to look at an activity and then to check your copy of the National Science Education Standards so that you can read the specific Content Standard and Guide to the Content Standard.

For example, part of the correlations for the first activity, Photographing a Crime Scene, look like this for grades 5–8:

Activity	Content standard	Bullet number	Content description	Bullet number(s)
1. Photographing a Crime Scene	A	2	Understandings about scientific inquiry	1–5

This indicates that under Content Standard A, you will see some bulleted items. The number 2 in the third column indicates that you need to count down to the second bullet. The second bulleted item says, "Understandings about scientific inquiry." If you look down under the section titled "Guide to the Content Standard," you will see that there is a section that has the exact same title, "Understandings about scientific inquiry." Under that section, you will find a number of bulleted items again. In this case, there are seven of them, and this activity is correlated to the first five.

National Science Education Standards Correlations (Grades 5–8)

Activity	Content standard	Bullet number	5–8 content description	Bullet number(s)
1. Photographing a Crime Scene	A	1	Abilities necessary to do scientific inquiry	1–7
	A	2	Understandings about scientific inquiry	1–5
	E	1	Abilities of technological design	1–5
	E	2	Understandings about science and technology	4–5
	G	1	Science as a human endeavor	1–2
2. Processing a Crime Scene	A	1	Abilities necessary to do scientific inquiry	1–7
	A	2	Understandings about scientific inquiry	1–5
	E	1	Abilities of technological design	1–5

(continued on next page)

Activity	Content standard	Bullet number	5–8 content description	Bullet number(s)
	E	2	Understandings about science and technology	4–5
	G	1	Science as a human endeavor	1–2
3. Fingerprints	A	1	Abilities necessary to do scientific inquiry	1–7
	A	2	Understandings about scientific inquiry	1–5
	E	1	Abilities of technological design	1–5
	E	2	Understandings about science and technology	4–5
	G	1	Science as a human endeavor	1–2
4. Detecting Blood	A	1	Abilities necessary to do scientific inquiry	1–7
	A	2	Understandings about scientific inquiry	1–5
	E	1	Abilities of technological design	1–5
	E	2	Understandings about science and technology	4–5
	G	1	Science as a human endeavor	1–2
5. Matching DNA	A	1	Abilities necessary to do scientific inquiry	1–7
	A	2	Understandings about scientific inquiry	1–5
	C	2	Reproduction and heredity	4–5
	C	5	Diversity and adaptations of organisms	1
	E	1	Abilities of technological design	1–5
	E	2	Understandings about science and technology	4–5
	G	1	Science as a human endeavor	1–2

(continued on next page)

Activity	Content standard	Bullet number	5–8 content description	Bullet number(s)
6. Blood Pattern Analysis	A	1	Abilities necessary to do scientific inquiry	1–7
	A	2	Understandings about scientific inquiry	1–5
	B	1	Properties and changes of properties in matter	1
	B	2	Motions and forces	1, 3
	E	1	Abilities of technological design	1–5
	E	2	Understandings about science and technology	4–5
	G	1	Science as a human endeavor	1–2
7. Car Accident	A	1	Abilities necessary to do scientific inquiry	1–7
	A	2	Understandings about scientific inquiry	1–5
	B	1	Properties and changes of properties in matter	1
	B	2	Motions and forces	1, 3
	E	1	Abilities of technological design	1–5
	E	2	Understandings about science and technology	4–5
	G	1	Science as a human endeavor	1–2
8. Tire Tracks	A	1	Abilities necessary to do scientific inquiry	1–7
	A	2	Understandings about scientific inquiry	1–5
	B	1	Properties and changes of properties in matter	1
	E	1	Abilities of technological design	1–5
	E	2	Understandings about science and technology	4–5
	G	1	Science as a human endeavor	1–2

(continued on next page)

Activity	Content standard	Bullet number	5–8 content description	Bullet number(s)
9. Density of Glass Fragments	A	1	Abilities necessary to do scientific inquiry	1–7
	A	2	Understandings about scientific inquiry	1–5
	B	1	Properties and changes of properties in matter	1
	E	1	Abilities of technological design	1–5
	E	2	Understandings about science and technology	4–5
	G	1	Science as a human endeavor	1–2
10. Glass Fracture Patterns	A	1	Abilities necessary to do scientific inquiry	1–7
	A	2	Understandings about scientific inquiry	1–5
	B	2	Motions and forces	1, 3
	E	1	Abilities of technological design	1–5
	E	2	Understandings about science and technology	4–5
	G	1	Science as a human endeavor	1–2
11. Physical Properties of Soil	A	1	Abilities necessary to do scientific inquiry	1–7
	A	2	Understandings about scientific inquiry	1–5
	B	1	Properties and changes of properties in matter	1
	D	1	Structure of the earth system	4–5
	E	1	Abilities of technological design	1–5
	E	2	Understandings about science and technology	4–5
	G	1	Science as a human endeavor	1–2

(continued on next page)

Activity	Content standard	Bullet number	5–8 content description	Bullet number(s)
12. Components of Soil	A	1	Abilities necessary to do scientific inquiry	1–7
	A	2	Understandings about scientific inquiry	1–5
	B	1	Properties and changes of properties in matter	1
	D	1	Structure of the earth system	4–5
	E	1	Abilities of technological design	1–5
	E	2	Understandings about science and technology	4–5
	G	1	Science as a human endeavor	1–2
13. Handwriting Analysis	A	1	Abilities necessary to do scientific inquiry	1–7
	A	2	Understandings about scientific inquiry	1–5
	E	1	Abilities of technological design	1–5
	E	2	Understandings about science and technology	4–5
	G	1	Science as a human endeavor	1–2
14. Searching Through Garbage	A	1	Abilities necessary to do scientific inquiry	1–7
	A	2	Understandings about scientific inquiry	1–5
	E	1	Abilities of technological design	1–5
	E	2	Understandings about science and technology	4–5
	G	1	Science as a human endeavor	1–2

(continued on next page)

Activity	Content standard	Bullet number	5–8 content description	Bullet number(s)
15. Shoe Prints	A	1	Abilities necessary to do scientific inquiry	1–7
	A	2	Understandings about scientific inquiry	1–5
	E	1	Abilities of technological design	1–5
	E	2	Understandings about science and technology	4–5
	G	1	Science as a human endeavor	1–2
16. Tool Marks	A	1	Abilities necessary to do scientific inquiry	1–7
	A	2	Understandings about scientific inquiry	1–5
	E	1	Abilities of technological design	1–5
	E	2	Understandings about science and technology	4–5
	G	1	Science as a human endeavor	1–2
17. Microscopic Fibers	A	1	Abilities necessary to do scientific inquiry	1–7
	A	2	Understandings about scientific inquiry	1–5
	E	1	Abilities of technological design	1–5
	E	2	Understandings about science and technology	4–5
	G	1	Science as a human endeavor	1–2

(continued on next page)

Activity	Content standard	Bullet number	5–8 content description	Bullet number(s)
18. Human Hair versus Animal Hair	A	1	Abilities necessary to do scientific inquiry	1–7
	A	2	Understandings about scientific inquiry	1–5
	C	2	Reproduction and heredity	4–5
	C	5	Diversity and adaptations of organisms	1
	E	1	Abilities of technological design	1–5
	E	2	Understandings about science and technology	4–5
	G	1	Science as a human endeavor	1–2
19. Summary: A Car As a Crime Scene	A	1	Abilities necessary to do scientific inquiry	1–7
	A	2	Understandings about scientific inquiry	1–5
	E	1	Abilities of technological design	1–5
	E	2	Understandings about science and technology	4–5
	G	1	Science as a human endeavor	1–2
20. Summary: Missing Person—Your Teacher!	A	1	Abilities necessary to do scientific inquiry	1–7
	A	2	Understandings about scientific inquiry	1–5
	E	1	Abilities of technological design	1–5
	E	2	Understandings about science and technology	4–5
	G	1	Science as a human endeavor	1–2

NATIONAL SCIENCE EDUCATION STANDARDS CORRELATIONS (GRADES 9–12)

Activity	Content standard	Bullet number	9–12 content description	Bullet number(s)
1. Photographing a Crime Scene	A	1	Abilities necessary to do scientific inquiry	1–5
	A	2	Understandings about scientific inquiry	3, 5
	E	1	Abilities of technological design	1–4
	G	1	Science as a human endeavor	2
2. Processing a Crime Scene	A	1	Abilities necessary to do scientific inquiry	1–5
	A	2	Understandings about scientific inquiry	3, 5
	E	1	Abilities of technological design	1–4
	G	1	Science as a human endeavor	2
3. Fingerprints	A	1	Abilities necessary to do scientific inquiry	1–5
	A	2	Understandings about scientific inquiry	3, 5
	B	3	Chemical reactions	1, 5
	E	1	Abilities of technological design	1–4
	G	1	Science as a human endeavor	2
4. Detecting Blood	A	1	Abilities necessary to do scientific inquiry	1–5
	A	2	Understandings about scientific inquiry	3, 5
	B	3	Chemical reactions	1, 5
	C	2	Molecular basis of heredity	1–2
	E	1	Abilities of technological design	1–4
	G	1	Science as a human endeavor	2

(continued on next page)

Activity	Content standard	Bullet number	9–12 content description	Bullet number(s)
5. Matching DNA	A	1	Abilities necessary to do scientific inquiry	1–5
	A	2	Understandings about scientific inquiry	3, 5
	C	2	Molecular basis of heredity	1–2
	E	1	Abilities of technological design	1–4
	G	1	Science as a human endeavor	2
6. Blood Pattern Analysis	A	1	Abilities necessary to do scientific inquiry	1–5
	A	2	Understandings about scientific inquiry	3, 5
	B	4	Motions and forces	1
	E	1	Abilities of technological design	1–4
	G	1	Science as a human endeavor	2
7. Car Accident	A	1	Abilities necessary to do scientific inquiry	1–5
	A	2	Understandings about scientific inquiry	3, 5
	B	4	Motions and forces	1
	E	1	Abilities of technological design	1–4
	G	1	Science as a human endeavor	2
8. Tire Tracks	A	1	Abilities necessary to do scientific inquiry	1–5
	A	2	Understandings about scientific inquiry	3, 5
	E	1	Abilities of technological design	1–4
	G	1	Science as a human endeavor	2

(continued on next page)

Activity	Content standard	Bullet number	9–12 content description	Bullet number(s)
9. Density of Glass Fragments	A	1	Abilities necessary to do scientific inquiry	1–5
	A	2	Understandings about scientific inquiry	3, 5
	B	2	Structure and properties of matter	4
	E	1	Abilities of technological design	1–4
	G	1	Science as a human endeavor	2
10. Glass Fracture Patterns	A	1	Abilities necessary to do scientific inquiry	1–5
	A	2	Understandings about scientific inquiry	3, 5
	B	2	Structure and properties of matter	4
	E	1	Abilities of technological design	1–4
	G	1	Science as a human endeavor	2
11. Physical Properties of Soil	A	1	Abilities necessary to do scientific inquiry	1–5
	A	2	Understandings about scientific inquiry	3, 5
	B	2	Structure and properties of matter	4
	D	2	Geochemical cycles	1–2
	E	1	Abilities of technological design	1–4
	G	1	Science as a human endeavor	2

(continued on next page)

Activity	Content standard	Bullet number	9–12 content description	Bullet number(s)
12. Components of Soil	A	1	Abilities necessary to do scientific inquiry	1–5
	A	2	Understandings about scientific inquiry	3, 5
	B	2	Structure and properties of matter	4
	D	2	Geochemical cycles	1–2
	E	1	Abilities of technological design	1–4
	G	1	Science as a human endeavor	2
13. Handwriting Analysis	A	1	Abilities necessary to do scientific inquiry	1–5
	A	2	Understandings about scientific inquiry	3, 5
	E	1	Abilities of technological design	1–4
	G	1	Science as a human endeavor	2
14. Searching Through Garbage	A	1	Abilities necessary to do scientific inquiry	1–5
	A	2	Understandings about scientific inquiry	3, 5
	E	1	Abilities of technological design	1–4
	G	1	Science as a human endeavor	2
15. Shoe Prints	A	1	Abilities necessary to do scientific inquiry	1–5
	A	2	Understandings about scientific inquiry	3, 5
	E	1	Abilities of technological design	1–4
	G	1	Science as a human endeavor	2

(continued on next page)

Activity	Content standard	Bullet number	9–12 content description	Bullet number(s)
16. Tool Marks	A	1	Abilities necessary to do scientific inquiry	1–5
	A	2	Understandings about scientific inquiry	3, 5
	E	1	Abilities of technological design	1–4
	G	1	Science as a human endeavor	2
17. Microscopic Fibers	A	1	Abilities necessary to do scientific inquiry	1–5
	A	2	Understandings about scientific inquiry	3, 5
	B	2	Structure and properties of matter	4
	E	1	Abilities of technological design	1–4
	G	1	Science as a human endeavor	2
18. Human Hair versus Animal Hair	A	1	Abilities necessary to do scientific inquiry	1–5
	A	2	Understandings about scientific inquiry	3, 5
	C	2	Molecular basis of heredity	1–2
	E	1	Abilities of technological design	1–4
	G	1	Science as a human endeavor	2
19. Summary: A Car As a Crime Scene	A	1	Abilities necessary to do scientific inquiry	1–5
	A	2	Understandings about scientific inquiry	3, 5
	B	4	Motions and forces	1
	E	1	Abilities of technological design	1–4
	E	2	Understandings about science and technology	3
	G	1	Science as a human endeavor	2

(continued on next page)

Activity	Content standard	Bullet number	9–12 content description	Bullet number(s)
20. Summary: Missing Person—Your Teacher!	A	1	Abilities necessary to do scientific inquiry	1–5
	A	2	Understandings about scientific inquiry	3, 5
	E	1	Abilities of technological design	1–4
	E	2	Understandings about science and technology	3
	G	1	Science as a human endeavor	2

General Rubric

Criteria	Procedures and reasoning	Strategies	Communication and use of data	Concepts and content
Level				
1 (Does not meet the standard)	• Did not use scientific procedures or tools to collect data	• Failed to use reasoning • Failed to use a strategy • Failed to use a procedure	• Data not recorded • Conclusion based on data not reached • Could not use scientific terms, symbols, graphs, and so forth • Explanation of task not given or was not connected to data	• No use, little use, or inappropriate use of scientific terms • No use, little use, or inappropriate use of scientific theories or principles • No understanding, little understanding, or inappropriate understanding of the various properties or materials used in task
2 (Partially meets the standard)	• Attempted to use scientific procedures or tools to collect data, but some collected data was inaccurate or incomplete	• Used reasoning, but only completed part of the task • Used a strategy, but was not effective in completing the task • Used a procedure, but could not collect data or form a conclusion	• Data recorded but not complete • Conclusion reached not fully supported by collected data • Attempted to use scientific terms, symbols, graphs, and so forth, but used incompletely and with missing components • Conclusions that were reached were not clear.	• Some use of appropriate scientific terms • Some use of appropriate scientific theories or principles • Some understanding of the various properties or materials used in the task
3 (Meets the standard)	• Used some scientific procedures or tools effectively to collect data with only minimal error	• Used effective reasoning • Used a strategy that allowed student to complete the task • Used a procedure, recorded data, conducted an experiment, and asked questions that could be tested	• Data recorded clearly • Conclusions essentially supported by collected data • Used scientific terms, symbols, graphs, and so forth • Conclusions that were reached were presented clearly.	• Appropriate use of scientific terms • Appropriate use of scientific theories or principles • Appropriate understanding of the various properties and materials used in the task

(continued on next page)

Criteria	Procedures and reasoning	Strategies	Communication and use of data	Concepts and content
4 (Exceeds the standard)	• Used scientific procedures and tools accurately to skillfully collect, analyze, and evaluate data	• Used advanced reasoning and connected observed effects with their causes • Designed a clear strategy, used the strategy, and adapted the strategy when necessary • Employed a procedure that took full advantage of all characteristics of the scientific method	• Data was recorded clearly and analyzed correctly. Further data was collected to clarify or to find error. • Conclusions were supported by collected data. Other questions and concepts that were suggested by the data were explored. • Scientific terms, symbols, graphs, and so forth were used. Multiple forms for presenting data were used. • The conclusions that were presented were clear. An effective presentation reviewing the task was used. All concepts were explained without need for further clarification.	• Complete and appropriate use of precise scientific terms; applied terms learned prior to activity from other activities • Complete and clear understanding of scientific theories or principles; referenced evidence in a relevant manner • Applied new knowledge to revise during process • Use of properties and materials suggested understanding beyond the scope of the activity. New questions related to material were formed.

1. Photographing a Crime Scene

TEACHER RESOURCE PAGE

INSTRUCTIONAL OBJECTIVES

Students will be able to:

- record the layout of a crime scene
- determine the important aspects of the scene
- evaluate the quality of photographs in re-creating a crime scene

NATIONAL SCIENCE EDUCATION STANDARDS CORRELATIONS

GRADES 5–8

Content standard	Bullet number	Content description	Bullet number(s)
A	1	Abilities necessary to do scientific inquiry	1–7
A	2	Understandings about scientific inquiry	1–5
E	1	Abilities of technological design	1–5
E	2	Understandings about science and technology	4–5
G	1	Science as a human endeavor	1–2

GRADES 9–12

Content standard	Bullet number	Content description	Bullet number(s)
A	1	Abilities necessary to do scientific inquiry	1–5
A	2	Understandings about scientific inquiry	3, 5
E	1	Abilities of technological design	1–4
G	1	Science as a human endeavor	2

VOCABULARY

- **close-up:** picture taken from a range that allows small details to be recorded, such as dust, fingerprints, or fibers
- **middle range:** the range in which parts of the crime scene are photographed in a way that captures the general location of items without focusing in on minute details
- **photography:** process for recording images of a crime scene using digital or film cameras

1. Photographing a Crime Scene

TEACHER RESOURCE PAGE

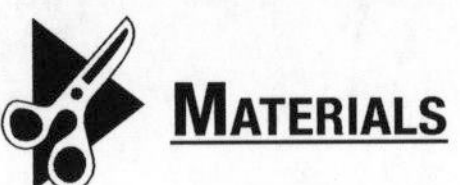

Materials

- digital camera
- computer
- crime scene
- paper

Helpful Hints and Discussion

Time frame: one class period
Structure: partners or groups
Location: various locations outside of the classroom

You will need to provide some locations where one student can take pictures while the other students wait. The locations can be made to look like a crime scene by carefully overturning furniture or by leaving "evidence" such as broken glass, fake blood, obvious fingerprints, fake weapons, tools, hair samples, and so forth. You might wish to acquire some crime-scene tape to mark the boundaries of the area where students will be taking pictures.

Meeting the Needs of Diverse Learners

You might find that students with different abilities will benefit from extra help or extra challenges. Students who need extra challenges should complete the Extension Option and the Follow-Up Activities. These students should also be encouraged to work with struggling classmates with the use of equipment and choosing important aspects of the crime scene. Students who need extra help should be encouraged to define the vocabulary words and to keep them in a word bank for later use. They should also be allowed to work with a partner if necessary.

Scoring Rubric

Students meet the standard for this activity by:

- correctly using the camera and computer to collect and manipulate photographs
- appropriately using overview, middle range, and close-up picture selection
- correctly using the terms *middle range* and *close-up* to describe images
- capturing a rough representation of the layout of the crime scene and evidence through photography

1. Photographing a Crime Scene

TEACHER RESOURCE PAGE

RECOMMENDED INTERNET SITES

- **Crime-Scene-Investigator.net—Crime Scene and Evidence Photography**
  www.crime-scene-investigator.net/csi-photo.html
- **PoliceOne.com—Crime scene imaging techniques**
  www.policeone.com/police-products/investigation/forensic-supplies/articles/86103-Crime-scene-imaging-techniques

ANSWER KEY

1. Answers will vary, but the number of pictures taken will depend on the nature of the crime, the nature of the scene, and the kind of evidence that is present.
2. Answers will vary, but in general it is difficult to re-create a crime scene from photographs unless careful notes are taken and measurements are made to place objects relative to some stationary object at the scene.
3. Precise protocol for taking pictures at a crime scene allows photographers to follow a checklist. Following a checklist increases the chance that photographers will not miss any important evidence.

Name ______________________ Date ______________________

1. Photographing a Crime Scene

STUDENT ACTIVITY PAGE

OBJECTIVE

To accurately record a crime scene using photography

BEFORE YOU BEGIN

Photography is one of the main ways that investigators can preserve a crime scene. Some crime scenes, such as a highway, cannot be left closed for months. As many details as possible must be recorded. These details might be used in the investigation and even the prosecution of a crime.

Photographers generally start by taking pictures of the entire crime scene in a way that shows the overall layout. These pictures should show the relationship between the various objects in the room or crime scene. The images should follow a pattern that would allow someone who could not physically enter the crime scene to have an idea of where all of the objects were relative to one another. Then photographers move into the **middle range** and capture the general arrangement of the various objects around the scene. What kind of books are on the shelves? Where is the body in relation to the murder weapon? Where do the car's skid marks start and finish?

The last thing a crime scene photographer must do is take many close-ups. **Close-ups** capture details of the scene. These details include things such as fingerprints, blood spatter, dust, fibers, alignment of the body, glass fragments, footprints, hair, clothing, wounds, tool marks, soil, stains, bite marks, and any other evidence that might be needed.

Photography is very useful for capturing images of items that might not survive to be collected. Examples include footprints in mud or snow, evidence that is being displaced by rain, or fingerprints that are too delicate to be collected mechanically.

MATERIALS

- digital camera
- computer
- crime scene
- paper

PROCEDURE

Note: Your teacher will split you into groups and will provide you with a location that will be your "crime scene." Only one member of the group will witness and take pictures of the scene with the digital camera.

1. Sketch a bird's-eye view of the crime scene before you take any pictures. Be sure to label the major objects and any important evidence.
2. Take ten overview pictures of the crime scene, numbering your pictures as you go. In the data table, write down why you took each picture and what it shows.
3. Take five pictures of the crime scene in middle range. Choose an area that seems important to the crime.
4. Take five close-ups of any evidence that is small or that might be damaged during collection.

Name ________________________________ Date ________________________________

1. Photographing a Crime Scene

STUDENT ACTIVITY PAGE

5. Transfer your pictures from the camera to a computer.
6. Have the members of your group who did not witness the crime scene firsthand view your pictures and sketch a blueprint-style layout of the crime scene as seen from a bird's-eye view.
7. Clearly mark the areas where evidence is located.

Extension Option

Find a photography resource, either a book or a reliable Web site, that explains the various skills needed to be a good photographer. Which skills would also be useful in crime-scene photography?

Data Collection and Analysis

Overview Pictures

Number of photograph	Reason taken	Important evidence shown
1.		
2.		
3.		
4.		
5.		

(chart continues on next page)

Name ______________________ Date ______________________

1. Photographing a Crime Scene

STUDENT ACTIVITY PAGE

OVERVIEW PICTURES (*CONTINUED*)

6.		
7.		
8.		
9.		
10.		

MID-RANGE PICTURES

Number of photograph	Reason taken	Important evidence shown
1.		
2.		

(chart continues on next page)

Name ______________________ Date ______________________

1. Photographing a Crime Scene

STUDENT ACTIVITY PAGE

MID-RANGE PICTURES (*CONTINUED*)

3.		
4.		
5.		

CLOSE-UP PICTURES

Number of photograph	Reason taken	Important evidence shown
1.		
2.		
3.		
4.		
5.		

Name ______________________ Date ______________________

1. Photographing a Crime Scene

STUDENT ACTIVITY PAGE

Concluding Questions

1. This activity only allows for 20 pictures in total. Did any part of the crime scene need more images taken? If so, explain which ones and why.

____

____

____

____

____

2. Compare the sketch drawn by the photographer and the sketch drawn by the group members who did not view the crime scene. What aspects of the scene were properly placed and what aspects were missing or misplaced?

____

____

____

____

____

3. How might having a plan for photographing a crime scene improve the overall value of the pictures that are taken?

____

____

____

____

____

Follow-Up Activities

1. Create a protocol for taking pictures of a crime scene.
2. Repeat the lab and have a different person operate the camera using the protocol you created.

2. Processing a Crime Scene

TEACHER RESOURCE PAGE

Instructional Objectives

Students will be able to:

- determine the best way to search a crime scene
- plan how and where they will collect evidence
- define the boundaries of a crime scene

National Science Education Standards Correlations

Grades 5–8

Content standard	Bullet number	Content description	Bullet number(s)
A	1	Abilities necessary to do scientific inquiry	1–7
A	2	Understandings about scientific inquiry	1–5
E	1	Abilities of technological design	1–5
E	2	Understandings about science and technology	4–5
G	1	Science as a human endeavor	1–2

Grades 9–12

Content standard	Bullet number	Content description	Bullet number(s)
A	1	Abilities necessary to do scientific inquiry	1–5
A	2	Understandings about scientific inquiry	3, 5
E	1	Abilities of technological design	1–4
G	1	Science as a human endeavor	2

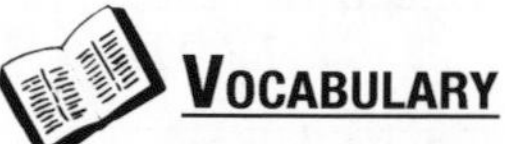

Vocabulary

- **fingertip search:** a search for evidence in which as much as the crime scene as possible is searched by hand, sometimes inch by inch
- **grid search:** a search for evidence in which the crime scene is broken up into sections, sometimes like squares on a checkerboard, and then searched systematically
- **macroscopic search:** a search for evidence that is generally large and easily seen with the naked eye
- **microscopic search:** a search for evidence that is not easily seen with the naked eye, such as fingerprints or DNA

- **primary crime scene:** the location where a crime was committed
- **processing:** the act of defining the boundaries of a crime scene and then searching within those boundaries to collect evidence that can be used in a court of law
- **secondary crime scene:** a location where evidence related to a crime that took place at another location might be found

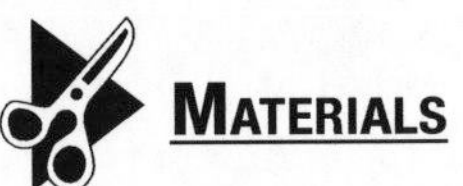

MATERIALS

- crime scene
- paper
- meterstick or tape measure
- gloves (latex and nitrile)

HELPFUL HINTS AND DISCUSSION

Time frame: one class period
Structure: partners or groups
Location: the classroom or another area that has been turned into a simulated crime scene

The locations can be made to look like crime scenes by carefully overturning furniture or by leaving "evidence" such as broken glass, fake blood, obvious fingerprints, fake weapons, notes, receipts, tools, hair samples, and so forth. Avoiding high-traffic areas in your school would be a good idea for this activity as students may choose to close off the crime scene. You might wish to acquire some crime-scene tape to mark the boundaries of the area where students will be working. Some students might be allergic to latex gloves, so be sure to have nitrile gloves available.

MEETING THE NEEDS OF DIVERSE LEARNERS

You might find that students with different abilities will benefit from extra help or extra challenges. Students who need extra challenges should complete the Extension Option and the Follow-Up Activity. These students could make good group leaders to help establish the boundaries of the crime scene. Students who need extra help might benefit from a review of the various kinds of evidence that might need to be collected from the crime scene. Emphasize that they need to be careful not to contaminate the crime scene with their fingerprints.

2. Processing a Crime Scene

TEACHER RESOURCE PAGE

SCORING RUBRIC

Students meet the standard for this activity by:

- correctly determining the boundaries and division of the crime scene using a grid
- collecting and organizing data on the evidence
- correctly using the grid in the search for evidence
- creating an accurate sketch of the crime scene and correctly recording the location of evidence

RECOMMENDED INTERNET SITES

- **Crime-Scene-Investigator.net—Crime Scene Response Guidelines**
  www.crime-scene-investigator.net/csi-response.html
- **International Crime Scene Investigators Association**
  www.icsia.org

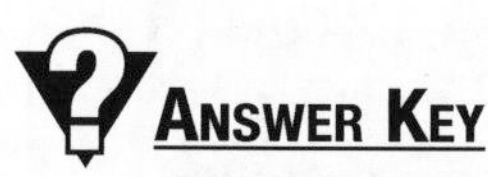

ANSWER KEY

1. Answers will vary, but students generally define the boundaries based on the location of obvious physical evidence. This means they will occasionally miss trace evidence.
2. Small evidence can be overlooked easily if the crime scene is not searched carefully.
3. It is unlikely that most crime scenes can be photographed from a bird's-eye view, so a sketch allows investigators to maintain a simple version of the location of objects and evidence. Photographs capture detail that could not be captured by a simple sketch.

Name ______________________ Date ______________________

2. Processing a Crime Scene

STUDENT ACTIVITY PAGE

OBJECTIVE

To determine the layout of a crime scene and the best way to search it for evidence

BEFORE YOU BEGIN

Police are usually first at the scene of a crime. They start the procedure of **processing** a crime scene. The police decide if a crime has been committed. They determine whether or not the crime is violent and if the suspect is still in the area. They must figure out if a weapon or some form of danger remains at the scene. If a victim is injured, the police must call for an ambulance and close off the scene.

At this point, a large number of people with a wide variety of jobs arrive. There are medical and fire responders, detectives, crime-scene investigators, medical examiners, and supervisors for all of these people. The scene may be a **primary crime scene** where the crime actually took place, or it could be a **secondary crime scene** where information or materials related to the crime have been found. In either type of scene, primary or secondary, investigators must process the area in a number of different ways. There is a **macroscopic (large) search,** in which obvious evidence is found (for example, a bloody knife or dead body). The **microscopic (small) search** is for objects that are not so easily seen or found. This might include things such as hair, DNA, fibers, chemicals, saliva, and things best observed with microscopes or by chemical treatment.

Different scenes are processed in different ways, but a couple of ways are common. In a **grid search,** the scene is split into sections, like the individual squares on a checkerboard, and then searched. Some searches, called **fingertip searches,** are done by hand. Sometimes these searches are done in full protective gear such as gloves, face masks, and coveralls. This protects the searchers and the evidence alike.

MATERIALS

- crime scene
- paper
- meterstick or tape measure
- gloves

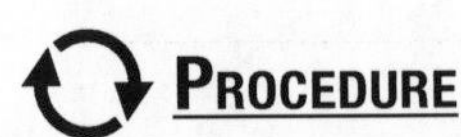

PROCEDURE

Note: Your teacher will split you into groups and will provide you with a location that will be your "crime scene." You will not be collecting any of the evidence you find other than to record what it is and where you found it.

1. Determine the boundaries of the crime scene that you are going to search.
2. Put on your gloves in case you have to move anything to look for evidence.
3. Determine what types of evidence are present and record them.
4. Write an overall description of the crime scene.
5. Make a detailed bird's-eye-view sketch of the crime scene. Choose a stationary object such as a large piece of furniture or a window and measure the distance from the object to some of the key pieces of evidence. Record the distances in the data table.

Name ______________________ Date ______________________

2. Processing a Crime Scene

STUDENT ACTIVITY PAGE

6. Make a list of photographs you would like taken by the crime-scene photographer. Your teacher may choose to take photographs if time permits.
7. Split the room into a grid. Each person should have a grid location.
8. Each person should search his or her part of the grid and record in detail any evidence found.

EXTENSION OPTION

Swap grids with another student and look for evidence again. Compare your findings and discuss the differences, if any, in your lists. Determine if a different type of search might have been more effective for this crime scene. Why or why not?

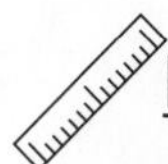

DATA COLLECTION AND ANALYSIS

Grid number	Evidence found	Distance from stationary reference

Name ______________________ Date ______________________

2. Processing a Crime Scene

STUDENT ACTIVITY PAGE

CONCLUDING QUESTIONS

1. Why did you choose the particular boundaries that you did for your crime scene?

2. What evidence was discovered in your grid search that might have been missed if only a general search had been done?

3. Why might a bird's-eye-view sketch be more useful than photographs? Why might photographs be better?

FOLLOW-UP ACTIVITY

Research the various legal problems a crime-scene investigator might face if he or she collected evidence without a search warrant. Contact your local police department and find out if the police would have needed a search warrant to collect evidence at your crime scene.

3. Fingerprints

TEACHER RESOURCE PAGE

INSTRUCTIONAL OBJECTIVES

Students will be able to:

- analyze the structure of a fingerprint
- determine if fingerprints match or not
- compare various fingerprints to identify similar structures

NATIONAL SCIENCE EDUCATION STANDARDS CORRELATIONS

GRADES 5–8

Content standard	Bullet number	Content description	Bullet number(s)
A	1	Abilities necessary to do scientific inquiry	1–7
A	2	Understandings about scientific inquiry	1–5
E	1	Abilities of technological design	1–5
E	2	Understandings about science and technology	4–5
G	1	Science as a human endeavor	1–2

GRADES 9–12

Content standard	Bullet number	Content description	Bullet number(s)
A	1	Abilities necessary to do scientific inquiry	1–5
A	2	Understandings about scientific inquiry	3, 5
B	3	Chemical reactions	1, 5
E	1	Abilities of technological design	1–4
G	1	Science as a human endeavor	2

VOCABULARY

- **arch:** a line that enters from one side of a fingerprint, follows a path that looks like a small hill, and then exits on the other side of the fingerprint
- **fingerprints:** patterns found in the skin on fingers that can be left on surfaces when those surfaces are handled by a person
- **invisible print:** a fingerprint usually made from the sweat and oils on fingers that is generally seen after dusting or other chemical treatment
- **loop:** a line that enters from one side of a fingerprint, curves around, and then exits on the same side
- **plastic** or **impression print:** a fingerprint that is usually found in a soft material

- **visible print:** a fingerprint made when material on the fingers is left behind
- **whorl:** a circular closed pattern in a fingerprint

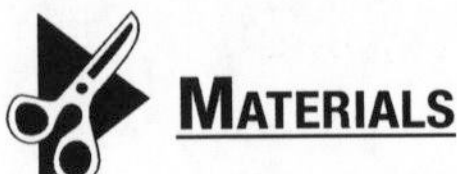

MATERIALS

- fingerprinting pad or pencil, paper, and clear tape (3/4-inch wide or wider)
- magnifying glass or microscope

HELPFUL HINTS AND DISCUSSION

Time frame: one class period
Structure: individuals or groups
Location: classroom

There are two different ways to do this activity. If you have equipment for fingerprinting, such as a traditional ink pad, you can take five index-finger fingerprints from a variety of students and then challenge students to identify structures such as loops and whorls. After that, you can randomize the samples and have students match them by finding specific structures that show that the two prints are from the same finger. You will need to take enough fingerprints so that each group can have five total prints with only two that match. You should assign a code to each person's fingerprints so that when you collect them, they can be shuffled and redistributed in a manner such that the teacher is sure which fingerprints match, but the students will be unable to tell without comparing them.

The other possibility, if you don't have a traditional ink pad, is to have students scribble on a piece of paper with a pencil until there is an area that is heavily covered in graphite. Students can then roll a finger across the graphite and then roll their finger across the sticky side of a piece of clear tape. The print will be lighter than with the ink-pad method, but the tape can be put on a piece of white paper to increase the visibility of the print.

MEETING THE NEEDS OF DIVERSE LEARNERS

You might find that students with different abilities will benefit from extra help or extra challenges. Students who need extra challenges should complete the Extension Option and the Follow-Up Activity. They can also be asked to take fingerprints from both index fingers and then determine if the prints are perfectly symmetrical or not. These students can also help peers who are struggling identify the various structures of the fingerprint. Encourage students who need extra help to look for labeled images on the Internet that they can use as a comparison when identifying structures of a fingerprint.

3. Fingerprints

SCORING RUBRIC

Students meet the standard for this activity by:

- correctly collecting fingerprints
- showing appropriate adaptation of the fingerprint-lifting process
- correctly using terms such as *loop, whorl,* and *arch*
- making the comparison of fingerprints by structure and not by "feel"

RECOMMENDED INTERNET SITES

- **Federal Bureau of Investigation—Taking Legible Fingerprints**
  www.fbi.gov/hq/cjisd/takingfps.html
- **Latent Print Examination**
  http://onin.com/fp

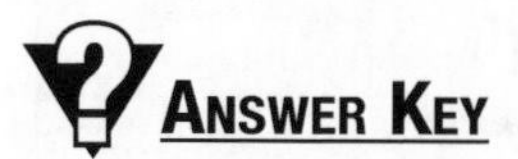

ANSWER KEY

1–2. Answers will vary depending on the quality of the prints.

3. Student opinions will vary, but students often cite whorls as the most obvious.

4. Answers will vary.

Name ______________________ Date ______________________

3. Fingerprints

STUDENT ACTIVITY PAGE

OBJECTIVE

To collect and compare fingerprints while looking for matching prints

BEFORE YOU BEGIN

If you look carefully at your fingertips, you will see a collection of lines that make up your **fingerprints.** Fingerprints have become widely accepted as evidence that is specific to only one person. Collecting and identifying fingerprints is an important part of forensics.

The lines on fingers have been classified into the categories of **loops, whorls,** and **arches.** A loop is a line that enters from one side of the fingerprint, curves around, and then exits from the same side. A whorl is a circular pattern that is closed. An arch is a line that enters from one side of the fingerprint, follows a path that looks like a small hill, and then exits on the other side. These are the basic descriptions. Professional fingerprint analyzers can describe the various kinds of loops, whorls, and arches with more detail and with subcategories.

Fingerprints on surfaces are also described in different ways. A **visible print** is usually made when material on the fingers, such as blood or dirt, is left behind. Visible prints can be seen easily. An **invisible print** is usually made from the sweat and oils that are on the fingers. These are generally seen after dusting or other chemical treatment. A **plastic** or **impression print** is found in a soft material. For example, if you grabbed onto something such as clay or wax, the material would take on the shape of the fingerprints.

MATERIALS

- fingerprinting pad or pencil, paper, and clear tape (3/4-inch wide or wider)
- magnifying glass or microscope

PROCEDURE

1. If you have a fingerprint pad, press your index finger firmly into the ink.
2. Roll the finger across the fingerprint chart on the next page, being careful not to drag it across the surface. Then answer the questions that follow.

   Note: If you do not have a fingerprint pad, an alternative way to collect fingerprints is to scribble heavily on paper with a pencil and then roll your index finger through the graphite. The fingerprint can then be collected directly onto a piece of clear tape. The tape can be put on a piece of white paper to increase the clarity of the print.
3. Repeat printing of the same finger until you have five clear samples.
4. Your teacher will collect the samples and mark them with numbers so that he or she can identify them after they have been shuffled.

Name ______________________ Date ______________________

3. Fingerprints

STUDENT ACTIVITY PAGE

5. Your teacher will give you a sample of five fingerprints. Two of them will match. Use a microscope or magnifying glass to identify loops, whorls, and arches.
6. Find the matching fingerprints, and indicate the structures that make you believe you have a match.

EXTENSION OPTION

Try to collect a fingerprint from an object. Rub your fingers over your skin to pick up some of the surface oils, perhaps from your forehead or cheeks, and press your fingers carefully to a glass beaker or a window. Try enhancing the fingerprint with different substances such as flour, iron filings, graphite dust, and any other material you might want to try. If you have access to a professional fingerprint-dusting kit, try that after you have experimented with the other possible materials. If you can enhance the print, try lifting it with a piece of tape.

DATA COLLECTION AND ANALYSIS

FINGERPRINT CHART

1. What, if any, motion caused your fingerprints to smear?

____

____

2. Is it clear that the five good fingerprint samples you took are all from the same finger?

____

____

Name ______________________ Date ______________________

3. Fingerprints

STUDENT ACTIVITY PAGE

Concluding Questions

1. What structures of the fingerprints were the easiest to identify?

2. Was there a structure of the fingerprints that was difficult to identify? Explain.

3. Which structures were the easiest to compare between two fingerprints?

4. When you determined that two fingerprints did not match, what structures were clearly the most different?

Follow-Up Activity

Use a scanner, photocopier, or digital camera to collect prints electronically. What advantages does this method of collection have over taking prints on paper or collecting them with tape?

4. Detecting Blood

TEACHER RESOURCE PAGE

INSTRUCTIONAL OBJECTIVES

Students will be able to:

- analyze a sample to see if it contains blood
- collect simulated blood samples
- draw conclusions about the nature of the sample they have collected

NATIONAL SCIENCE EDUCATION STANDARDS CORRELATIONS

GRADES 5–8

Content standard	Bullet number	Content description	Bullet number(s)
A	1	Abilities necessary to do scientific inquiry	1–7
A	2	Understandings about scientific inquiry	1–5
E	1	Abilities of technological design	1–5
E	2	Understandings about science and technology	4–5
G	1	Science as a human endeavor	1–2

GRADES 9–12

Content standard	Bullet number	Content description	Bullet number(s)
A	1	Abilities necessary to do scientific inquiry	1–5
A	2	Understandings about scientific inquiry	3, 5
B	3	Chemical reactions	1, 5
C	2	Molecular basis of heredity	1–2
E	1	Abilities of technological design	1–4
G	1	Science as a human endeavor	2

VOCABULARY

- **ABO system:** one of several systems for determining blood types in humans
- **blood:** a fluid found in humans that carries information, such as DNA, that is specific to an individual
- **blood type:** a way of classifying blood based on the various chemicals it might contain
- **chemiluminescence:** a chemical reaction that gives off light
- **Fluorescin:** a chemical used to reveal the presence of blood evidence
- **luminol:** a chemical used to reveal the presence of blood evidence

4. Detecting Blood

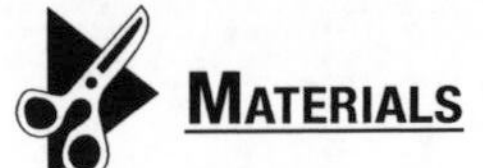

Materials

- artificial blood
- sterile swabs
- distilled water
- rinse bottles
- small beakers, 50- or 100-ml
- luminol
- potassium hydroxide
- hydrogen peroxide (3% over-the-counter concentration)
- coffee
- cola
- cranberry juice
- red food coloring
- old clothing or rags
- gloves (latex and nitrile)
- goggles
- color chart (found at most paint-supply stores)

Helpful Hints and Discussion

Time frame: one class period
Structure: individuals or groups
Location: the classroom and an area that can be made dark, such as a closet or storeroom

Most school districts do not allow students to handle real blood samples of any kind. Be sure to use artificial blood. If your school allows blood testing, be sure to request a copy of the guidelines and apply them to this activity.

The following Web sites both sell artificial blood that contains no blood products but still reacts with luminol:

- **Evident Crime Scene Products**
  www.evidentcrimescene.com
- **Crime Scene**
  www.crimescene.com

You will need to pre-stain some used clothing, rags, or surfaces with artificial blood, cola, cranberry juice, red food coloring, and coffee and allow them time to dry before the lab.

Some students might be allergic to latex gloves, so be sure to have nitrile gloves available.

Follow the steps below to make a stock luminol solution. **Be sure to wear goggles and gloves while mixing this solution.**

1. Weigh out 8 grams of luminol.
2. Weigh out 60 grams of potassium hydroxide. **Special care should be used when handling potassium hydroxide, as it is corrosive.**
3. Combine the two chemicals above in a 1,000-ml Erlenmeyer flask with 1,000 ml of distilled water. Stir or swirl gently until both of the solids have dissolved.
4. The finished luminol solution will also need to be handled with care since it contains the corrosive potassium hydroxide.

MEETING THE NEEDS OF DIVERSE LEARNERS

You might find that students with different abilities will benefit from extra help or extra challenges. Students who need extra challenges should complete the Extension Option and the Follow-Up Activity. These students can also help their peers with the safe handling of chemicals and evidence samples. Students who need extra help should be given careful instruction in all steps of the collecting and testing of the samples. Special attention should be given to the handling of chemicals and samples of unknown materials.

SCORING RUBRIC

Students meet the standard for this activity by:

- using the correct procedure for collecting and testing apparent blood samples
- recording data that indicates or does not indicate the presence of blood
- correctly using scientific terms such as *luminol* and *chemiluminescence*
- basing conclusions on evidence

RECOMMENDED INTERNET SITES

- **Crime-Scene-Investigator.net—Collection and Preservation of Blood Evidence from Crime Scenes**
  www.crime-scene-investigator.net/blood.html
- **Suite101.com: Luminol—Chemiluminescent Blood Detector**
  http://crime-scene-processing.suite101.com/article.cfm/chemiluminescent_luminol

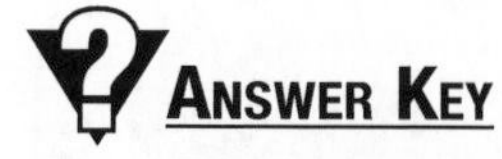

ANSWER KEY

1. Most students have seen dried blood and are pretty good at recognizing it when they see it. Depending on the quality of the artificial blood, they may think that it is not blood.
2. Students frequently notice the smell of coffee. Answers will vary.
3. The luminol test requires a dark workplace, and it can be difficult to do at a crime scene. This is why many samples aren't tested until they are back at a lab.

Name ______________________ Date ______________

4. Detecting Blood

STUDENT ACTIVITY PAGE

OBJECTIVE

To determine whether or not a collected sample is blood

BEFORE YOU BEGIN

The human body has 5 to 6 liters of **blood** in it. Blood can tell quite a story when it is outside the body. It can be found at a crime scene in a number of ways. Blood can appear as drops or pools that can be seen with the naked eye. Sometimes chemical agents are required to reveal the presence of blood evidence, especially if someone has tried to cover it up. You can spray **luminol** solution on a location where you suspect blood was spilled and then cleaned. If blood residue is present, the blood and luminol will react to glow light blue in the dark. This chemical reaction that produces light is called **chemiluminescence. Fluorescin** is another chemical that is far more sensitive than luminol. However, it has the disadvantage that it can only indicate the presence of blood when the sample is viewed under ultraviolet light.

More of the story of your blood can be found through blood typing. There are a number of ways that blood is classified. The **ABO system** views human blood as having four basic **blood types:** A, B, O, and AB. There are other systems that recognize far more individual parts of blood that would account for almost 600 testable components. While blood typing is fast being replaced with DNA matching, blood-typing tests are relatively inexpensive and quick. They can be used to determine if two blood samples are a general match or not.

MATERIALS

- sterile swabs
- distilled water
- rinse bottle
- small beaker, 50- or 100-ml
- luminol solution
- hydrogen peroxide
- gloves
- goggles
- color chart

PROCEDURE

1. Put on your gloves and goggles before you handle the unknown samples and testing chemicals. This protects both you and the samples.
2. Write a description of each stain in your data table before you disturb it. Indicate the color and the size. Find the closest similar color on the color chart and record it in the data table.
3. Wet your sterile swab with distilled water from the rinse bottle.
4. Wipe the swab through one stain.

Name ______________________ Date ______________________

4. Detecting Blood

5. Mix 10 ml of luminol solution and 10 ml of hydrogen peroxide in a small beaker. **Be careful handling the luminol solution as it contains corrosive chemicals. If you get it on your skin, tell your teacher and wash the chemicals off from your skin immediately.**
6. Take your sample and swab to the dark testing area and dip the swab into the solution. A blue glow indicates the presence of blood.
7. Be sure to number your results and to record whether or not there was a reaction for each sample.
8. Repeat steps 2 through 6 for the other samples. **Be sure to use a new swab and new luminol and hydrogen peroxide solution each time. Use a clean beaker for each sample.** Reusing the swab, the luminol solution, or a dirty beaker will contaminate your results.

EXTENSION OPTION

You have more than the sense of sight to rely on when collecting data. With your teacher's permission, smell each sample and see if you can identify it by smell. Remember, this is not always a safe way to test a sample, but some materials have an obvious odor that you can smell whether you want to or not.

DATA COLLECTION AND ANALYSIS

Sample number	Where collected	Description (color, size, shape)	Luminol test (positive or negative)

Name ______________________ Date ______________________

4. Detecting Blood

STUDENT ACTIVITY PAGE

CONCLUDING QUESTIONS

1. Which sample did you think was blood before it was tested? Why did you think so?

____

____

____

____

2. Did you notice any odors from the samples before you tested them? If so, what did they smell like?

____

____

____

____

3. What part of this test, if any, would be difficult to carry out during the daytime in an outside crime scene?

____

____

____

____

FOLLOW-UP ACTIVITY

Human blood and animal blood looks very much the same. Find out at least two ways that scientists can tell the difference and what methods they use to determine the difference.

5. Matching DNA

INSTRUCTIONAL OBJECTIVES

Students will be able to:

- compare DNA profiles
- explain how profiles are similar or different

NATIONAL SCIENCE EDUCATION STANDARDS CORRELATIONS

GRADES 5–8

Content standard	Bullet number	Content description	Bullet number(s)
A	1	Abilities necessary to do scientific inquiry	1–7
A	2	Understandings about scientific inquiry	1–5
C	2	Reproduction and heredity	4–5
C	5	Diversity and adaptations of organisms	1
E	1	Abilities of technological design	1–5
E	2	Understandings about science and technology	4–5
G	1	Science as a human endeavor	1–2

GRADES 9–12

Content standard	Bullet number	Content description	Bullet number(s)
A	1	Abilities necessary to do scientific inquiry	1–5
A	2	Understandings about scientific inquiry	3, 5
C	2	Molecular basis of heredity	1–2
E	1	Abilities of technological design	1–4
G	1	Science as a human endeavor	2

VOCABULARY

- **DNA:** the abbreviation for deoxyribonucleic acid, a molecule found in almost all living organisms
- **VNTRs:** abbreviation for variable number tandem repeats, a way of saying that there are short stretches of the DNA molecule that have identically repeating parts

5. Matching DNA

TEACHER RESOURCE PAGE

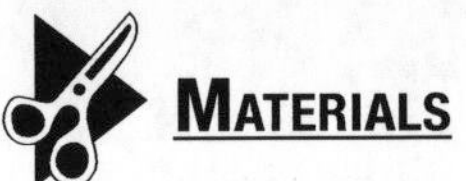

MATERIALS

- sample DNA profiles
- ruler

HELPFUL HINTS AND DISCUSSION

Time frame: 20 to 30 minutes
Structure: individuals
Location: classroom

This activity provides some simple DNA profiles for comparison and allows students to connect the idea that certain genetic features are handed down through generations. It also gives students the ability to match basic components of a DNA profile. There are five children in the example, three who are the children of the mother and father, one who has a different father (without the knowledge of the man who assumes he is the father), and one who was actually kidnapped. The back story that you may share with students is that the parents have been arrested because of their link to a crime, and authorities are unable to find documentation or birth records for the five children. The results provided in the activity are the sort that would be used to confirm that the man and woman are (or are not) the biological parents.

MEETING THE NEEDS OF DIVERSE LEARNERS

You might find that students with different abilities will benefit from extra help or extra challenges. Students who need extra challenges should complete the Extension Option and the Follow-Up Activity. These students should be able to explain in detail to their peers the parentage of all five children in the activity. Students who need extra help should be asked to find the three children who are clearly offspring of this mother and father based on their very similar DNA profiles.

SCORING RUBRIC

Students meet the standard for this activity by:

- recording data indicating which individuals share common genetic factors
- appropriately matching genetic factors
- correctly supporting conclusions of parentage with evidence
- correctly using and interpreting the charts indicating genetic factors

5. Matching DNA

TEACHER RESOURCE PAGE

Recommended Internet Sites

- **Oak Ridge National Laboratory—Human Genome Project Information: DNA Forensics**
  www.ornl.gov/sci/techresources/Human_Genome/elsi/forensics.shtml
- **ThinkQuest—Forensic DNA**
  http://library.thinkquest.org/28599/courtroom.htm

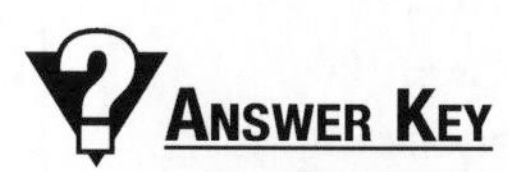

Answer Key

1.

Child	Child of this father	Child of this mother	Child of both parents	Child of one parent	Child of neither parent
Son 1	yes	yes	yes	no	no
Daughter 1	no	yes	no	yes	no
Son 2	yes	yes	yes	no	no
Daughter 2	yes	yes	yes	no	no
Daughter 3	no	no	no	no	yes

2. Daughter 3 is not the child of either the mother or the father. She was most likely kidnapped.
3. Daughter 1 is not the daughter of the father.
4. DNA evidence at a crime scene and DNA collected from a suspect can be matched much the same way except that all of the lines would match.

Name ______________________ Date ______________________

5. Matching DNA

STUDENT ACTIVITY PAGE

OBJECTIVE

To compare DNA profiles for similarities and differences

BEFORE YOU BEGIN

DNA stands for deoxyribonucleic acid. DNA is a molecule found in almost all living organisms. While most of the DNA found in two people is very similar, about one-tenth of 1 percent is different. This means that there are about 3 million molecular components that can be arranged in various ways to make genetically unique human beings.

At a crime scene, evidence might be collected that contains DNA. The DNA can be compared to samples collected from victims or criminals. DNA is found in blood, skin, hair, saliva, and other bodily fluids. DNA evidence must be collected carefully so that it is not contaminated. It must be treated carefully in the lab so that as much of the material as possible survives for testing.

VNTRs stands for variable number tandem repeats. This is just a way of saying that there are short stretches of the DNA molecule that have identically repeating parts. The number of times these parts repeat is generally different from person to person. These VNTRs make it possible for a properly trained technician to compare a relatively small number of markers in a short amount of time. Looking at the whole DNA structure takes much longer.

Once the data is collected, a DNA fingerprint is made. A DNA fingerprint can look like a series of bars on a chart or like sharp peaks on a line graph. Whatever the final presentation, a trained person can compare and identify identical (or different) DNA fingerprints just by looking at them.

MATERIALS

- ruler
- sample DNA profiles (see Procedure below)

PROCEDURE

1. Each numbered line represents a genetic trait, such as curly hair. Lines that are on the same level show that the child probably inherited that trait from that parent. In the example below, we see that the first son (S1) inherited trait 1 from his father because the markers in their DNA profiles are aligned. Repeat this process on the next page by comparing each parent's traits to those of each child (S1 = son 1, D1 = daughter 1, and so forth). Record your findings in the data tables by writing *yes* or *no* in each box.

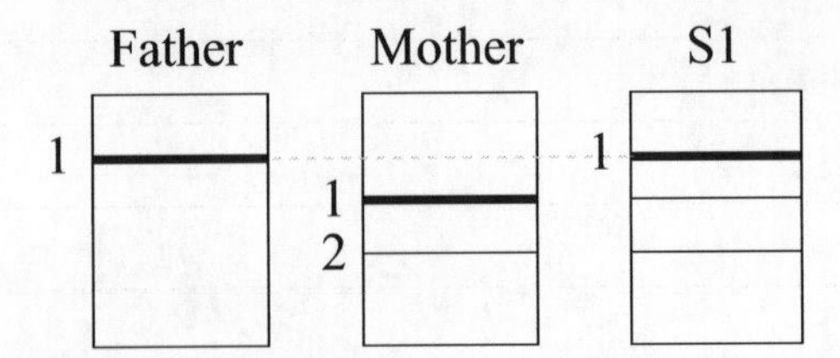

Name ______________________________ Date ______________________________

5. Matching DNA

STUDENT ACTIVITY PAGE

2. Check each child for trait 2, trait 3, and trait 4 of the father. Repeat with the traits for the mother.

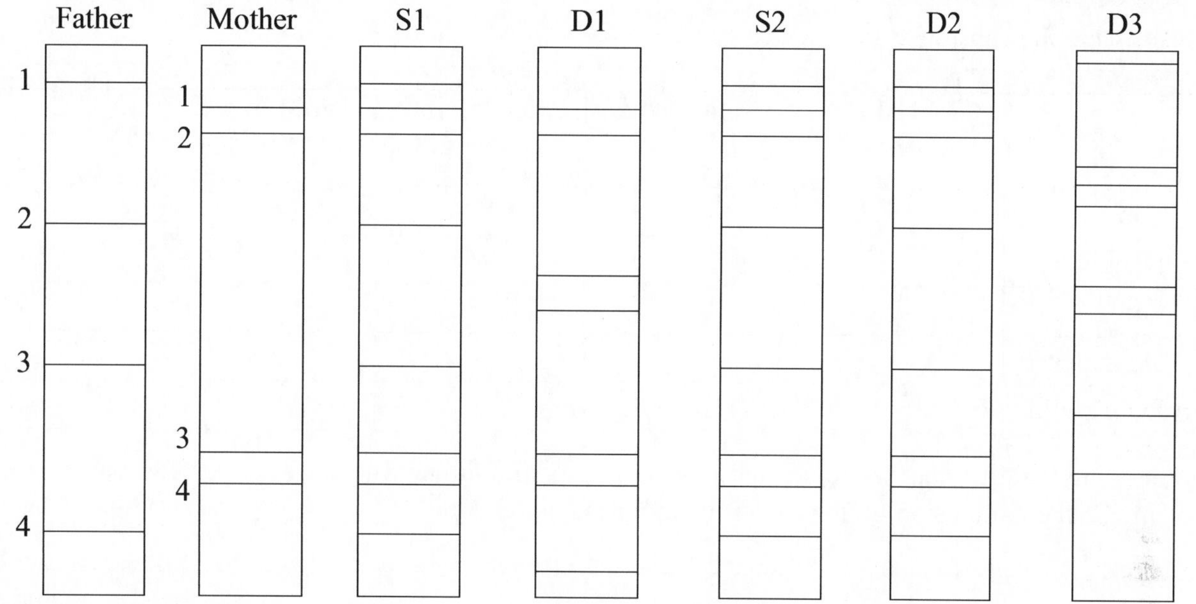

EXTENSION OPTION

Create two plausible DNA profiles for the mother and father of Daughter 3.

DATA COLLECTION AND ANALYSIS

Child	Shares trait 1 with father	Shares trait 2 with father	Shares trait 3 with father	Shares trait 4 with father	Child of this father
Son 1					
Daughter 1					
Son 2					
Daughter 2					
Daughter 3					

Child	Shares trait 1 with father	Shares trait 2 with father	Shares trait 3 with father	Shares trait 4 with father	Child of this father
Son 1					
Daughter 1					
Son 2					
Daughter 2					
Daughter 3					

Name ______________________________ Date ______________________________

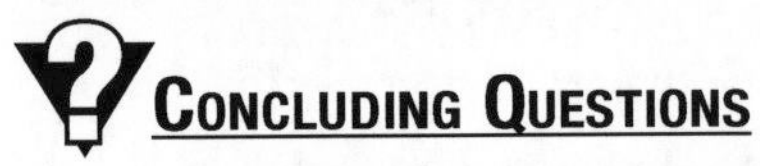

5. Matching DNA

STUDENT ACTIVITY PAGE

CONCLUDING QUESTIONS

1. Complete the chart below.

Child	Child of this father	Child of this mother	Child of both parents	Child of one parent	Child of neither parent
Son 1					
Daughter 1					
Son 2					
Daughter 2					
Daughter 3					

2. Knowing that this data was collected from a mother and father who were suspected of at least one crime, what conclusion might you draw about Daughter 3?

____

____

____

____

3. What might you suspect about Daughter 1?

____

____

____

____

4. How might this method be used to link a criminal to a crime using his or her DNA?

____

____

____

____

FOLLOW-UP ACTIVITY

Find some photographs of real DNA profiles upon which the simple profiles in this activity were based. Find out how many of the markers have to be in common between two samples before an investigator would be sure that he or she had a match.

6. Blood Pattern Analysis

TEACHER RESOURCE PAGE

Instructional Objectives

Students will be able to:

- define kinds of blood patterns they see
- describe blood patterns with correct terminology
- establish the manner in which a blood pattern was formed

National Science Education Standards Correlations

Grades 5–8

Content standard	Bullet number	Content description	Bullet number(s)
A	1	Abilities necessary to do scientific inquiry	1–7
A	2	Understandings about scientific inquiry	1–5
B	1	Properties and changes of properties in matter	1
B	2	Motions and forces	1, 3
E	1	Abilities of technological design	1–5
E	2	Understandings about science and technology	4–5
G	1	Science as a human endeavor	1–2

Grades 9–12

Content standard	Bullet number	Content description	Bullet number(s)
A	1	Abilities necessary to do scientific inquiry	1–5
A	2	Understandings about scientific inquiry	3, 5
B	4	Motions and forces	1
E	1	Abilities of technological design	1–4
G	1	Science as a human endeavor	2

Vocabulary

- **blood circles:** small circles formed when drops of blood impact a surface
- **blood pool:** blood that has collected in a puddle
- **blood smear:** blood pattern that shows that when two objects came into contact, one or both of them were covered in blood

- **blood splash:** a very distinct blood pattern, something like an exclamation point, that occurs when blood hits a surface at an angle of 30° or less
- **blood spurt:** a blood pattern formed when blood squirts onto a surface from a severed artery
- **blood trail:** larger blood smear left on the floor, wall, or other stationary object that reveals the direction a body was moving
- **crenellated blood:** a blood pattern that indicates that blood impacted a surface at a high speed
- **elliptical drops:** blood patterns that show the direction a blood source was moving and indicate that the blood landed while moving at an angle

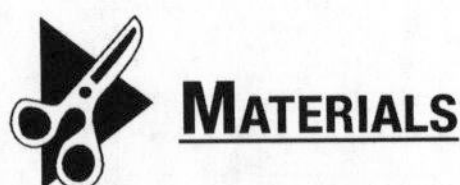

Materials

- artificial blood
- eyedropper
- rinse bottle
- protractor
- meterstick
- goggles
- apron
- paper
- tarp, drop cloth, or other protective sheet

Helpful Hints and Discussion

Time frame: one class period
Structure: individuals or partners
Location: classroom

Students will be creating blood stains using artificial blood or a substitute of similar viscosity. This activity can be messy, and students should wear goggles and aprons to protect their eyes and clothes. You might also wish to have designated areas for testing that are well protected with tarps or inexpensive plastic drop cloths from a paint-supply store. If you do the first Extension Option, be sure to warn students about the safety issues of working from a height.

Meeting the Needs of Diverse Learners

You might find that students with different abilities will benefit from extra help or extra challenges. Students who need extra challenges should complete the Extension Options and the Follow-Up Activity. These students should also be allowed to make other kinds of contact blood stains, such as blood trails. Students who need extra help might need assistance adjusting the angle of the paper that will be used for the landing surface for the blood.

6. Blood Pattern Analysis

SCORING RUBRIC

Students meet the standard for this activity by:

- following the procedure for creating a variety of patterns
- correctly recording the shapes created by the spattering process
- appropriately using terms to describe the blood-spatter patterns
- creating correct blood-spatter patterns

RECOMMENDED INTERNET SITES

- **Bloodstain Pattern Analysis Tutorial**
  www.bloodspatter.com/BPATutorial.htm
- **Brazoria County Sheriff's Office—Blood Spatter Interpretation**
  http://brazoria-county.com/sheriff/id/blood/index.htm

ANSWER KEY

1. The drops became more elongated as the angle increased.
2. The drops became somewhat narrower as the angle increased.
3. As the height increased, the drops generally became larger.
4. As the height increased, the drops generally became wider.
5. As the speed of the spurt increased, the spatter became more angular and elongated in the direction of travel.

Name ______________________________ Date ______________________________

6. Blood Pattern Analysis

STUDENT ACTIVITY PAGE

OBJECTIVE

To correctly observe and identify kinds of blood patterns

BEFORE YOU BEGIN

The way that blood forms patterns at a crime scene can tell investigators a lot about the crime that took place. **Blood pools** tell investigators that victims were not killed rapidly, since death causes the flow of blood to stop. Perfect **bloody circles** show that the blood was moving relatively slowly and perpendicularly to the surface it impacted. **Elliptical drops** show that the blood landed while moving at an angle. (Imagine a bloody suspect running from the scene of a murder.) They also show the direction the blood source was moving. **Blood smears** show that two objects came into contact and that one or both were covered in blood. This could mean that a clean object passed through a blood stain or a bloody object moved across a clean surface.

Crenellated blood marks are found when blood impacts a surface from a great height or at a high speed. **Blood spurt,** such as from a severed artery, leaves a very distinct pattern on a surface. It is easily identified. A **blood splash** usually has a very distinct shape, something like an exclamation point. It occurs when blood hits a surface at an angle of 30° or less. **Blood trails** are generally larger blood smears left on the floor, wall, or other stationary object that reveal the direction a body was moving. There are far more classifications of blood patterns, but even these few can reveal a lot to a trained forensic expert.

MATERIALS

- artificial blood
- eyedropper
- rinse bottle
- protractor
- meterstick
- goggles
- apron
- paper
- tarp, drop cloth, or other protective sheet

Name ______________________________ Date ______________________________

6. Blood Pattern Analysis

STUDENT ACTIVITY PAGE

PROCEDURE

1. Put on your goggles and apron.
2. Fill an eyedropper with artificial blood.
3. Let one drop fall on a flat sheet of paper (0° above the horizontal) from a height of 30 centimeters (about 1 foot).
4. Draw a sketch of the drop in your data table. The sketch does not have to be full-sized.
5. Measure the widest part of the drop and record the width in your data table.
6. Tilt the sheet of paper to 15° above the horizontal (tape it to a book or board if necessary).
7. Let a drop fall from 30 centimeters. Measure the widest part of the drop and record the width in your data table. Sketch the drop in your data table. The sketch does not have to be full-sized.
8. Repeat the dropping procedure for 30°, 45°, 60°, and 75°. Be sure to sketch each one and to measure the widest part of the drop so the width can be recorded in your data table.
9. Repeat the dropping procedure for 0°, 15°, 30°, 45°, 60°, and 75° from a height of 1 meter.
10. Your teacher will have a tarp or waterproof sheet on one wall of your classroom. Tape a sheet of paper to the sheet and squirt blood across it from the rinse bottle in a slow motion from left to right. Make a sketch of blood spurt in your data table.
11. Repeat the blood spurt at a medium speed and fast speed. Sketch the results in your data table.

EXTENSION OPTIONS

- Repeat the dropping procedure for 0°, 15°, 30°, 45°, 60°, and 75° from a second-story window or from a location provided by your teacher.
- Create a drop from a "mystery angle" and have other students try to determine which angle produces a spatter mark that best matches the "mystery angle."

Name ______________________ Date ______________________

6. Blood Pattern Analysis

STUDENT ACTIVITY PAGE

DATA COLLECTION AND ANALYSIS

Angle	Height	Sketch	Width
0°	30 cm		
15°	30 cm		
30°	30 cm		
45°	30 cm		
60°	30 cm		
75°	30 cm		

Name ______________________ Date ______________________

6. Blood Pattern Analysis

STUDENT ACTIVITY PAGE

Angle	Height	Sketch	Width
0°	1 m		
15°	1 m		
30°	1 m		
45°	1 m		
60°	1 m		
75°	1 m		

Name ______________________ Date ______________________

6. Blood Pattern Analysis

STUDENT ACTIVITY PAGE

Extension angle	Height	Sketch	Width
0°			
15°			
30°			
45°			
60°			
75°			

Name ______________________ Date ______________________

6. Blood Pattern Analysis

STUDENT ACTIVITY PAGE

Speed of blood spurt	Sketch
Slow	
Medium	
Fast	

CONCLUDING QUESTIONS

1. How did changing the angle at which the drop hit the paper affect the shape?

____

____

____

____

2. How did changing the angle at which the drop hit the paper affect the width?

____

____

____

____

6. Blood Pattern Analysis

3. How did changing the height from which the drop started affect the shape of the drop?

____

____

____

____

4. How did changing the height from which the drop started affect the width of the drop?

____

____

____

____

5. How did changing the speed at which the blood spurt hit the paper on the wall change the shape of the spatter on the paper?

____

____

____

____

FOLLOW-UP ACTIVITY

Research blood-spatter patterns and make a list of all the terms that are used to describe the various shapes of blood evidence found at crime scenes.

7. Car Accident

TEACHER RESOURCE PAGE

INSTRUCTIONAL OBJECTIVES

Students will be able to:

- calculate the speed of a car from the length of skid marks it leaves
- explain factors that affect skid mark length
- evaluate factors that affect skid reconstructions

NATIONAL SCIENCE EDUCATION STANDARDS CORRELATIONS

GRADES 5–8

Content standard	Bullet number	Content description	Bullet number(s)
A	1	Abilities necessary to do scientific inquiry	1–7
A	2	Understandings about scientific inquiry	1–5
B	1	Properties and changes of properties in matter	1
B	2	Motions and forces	1, 3
E	1	Abilities of technological design	1–5
E	2	Understandings about science and technology	4–5
G	1	Science as a human endeavor	1–2

GRADES 9–12

Content standard	Bullet number	Content description	Bullet number(s)
A	1	Abilities necessary to do scientific inquiry	1–5
A	2	Understandings about scientific inquiry	3, 5
B	4	Motions and forces	1
E	1	Abilities of technological design	1–4
G	1	Science as a human endeavor	2

VOCABULARY

- **coefficient of friction:** a comparison of the amount of friction experienced between two surfaces in contact
- **skid formula:** a formula used to estimate the distance a car traveled after brakes were applied based on the length of skid marks left by the tires

7. Car Accident

TEACHER RESOURCE PAGE

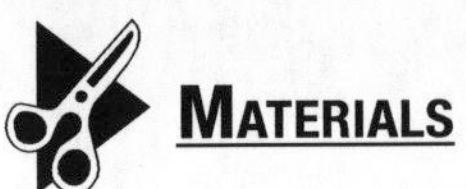

MATERIALS

- skid marks to measure (real or created)
- tape measure

HELPFUL HINTS AND DISCUSSION

Time frame: 30 minutes
Structure: partners
Location: classroom or other location

Students will measure distances to use in the skid formula. Some of the variables will be provided for them. Since real skid marks may be unavailable, you might try measuring the length of the classroom, part of a hallway, the length of the gym floor, a section of the parking lot, and so forth. It would be best if students had at least four distances of widely differing lengths to measure, especially something very long and very short. Students will determine how far a car was traveling if it skidded the distance they measured. If you can find real skid marks to measure, all the better, but **be sure that students are safe from traffic while making measurements.**

MEETING THE NEEDS OF DIVERSE LEARNERS

You might find that students with different abilities will benefit from extra help or extra challenges. Students who need extra challenges should complete the Extension Option and the Follow-Up Activity. These students can find more coefficients of friction for varying surfaces and make calculations for different circumstances. Students who need extra help might need assistance with the calculation and further explanation about the coefficient of friction.

SCORING RUBRIC

Students meet the standard for this activity by:

- correctly recording lengths of skid marks
- producing velocities of cars using the skid formula as a procedure
- applying the formula in more than one context by suggesting how road conditions might affect results

RECOMMENDED INTERNET SITES

- **Car Accidents.com—Accident Reconstruction**
  www.car-accidents.com/pages/accident_reconstruction.html
- **Walters Forensic Engineering—Skidmark Analysis & Braking**
  www.waltersforensic.com/articles/accident_reconstruction/vol1-no8.htm

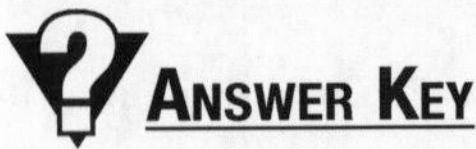

ANSWER KEY

1. No. It increases by a factor of the square root of 2.
2. This formula assumes the vehicle skidded to a stop with no collisions.
3. A different coefficient of friction is needed for each different surface.

Name ______________________ Date ______________________

7. Car Accident

STUDENT ACTIVITY PAGE

Objective

To use skid marks to determine the speed of a car in an accident

Before You Begin

Cars are often involved in accidents and crimes. Accidents can be caused by factors such as driver inattention, road conditions, and weather. The end result is often damage to the driver's vehicle, people, property, or other vehicles. The nature of that damage can often be used to help investigators determine how the damage was caused and whether or not it was by accident.

After an accident, emergency responders first make sure that injured people are treated or made safe. While receiving medical treatment, people involved in the accident can expect to be tested to see if they have drugs or alcohol in their systems. Car accidents often take place in the presence of witnesses. It is important for police to take the statements of as many people as possible when reconstructing an accident. People might not remember exactly what they saw. However, if a large number of people remember seeing a car run a red light, then that is probably what happened. Police must also take careful inventory of the inside of each vehicle in an accident. They have to keep track of tire tracks or collision marks around the accident. They must record damage to the vehicle. They should also try to keep the car as pristine as possible in case it has to be moved.

Skid marks can be measured and formulas such as the **skid formula** can be used to find out how fast a car was going when the accident started. One skid formula is $V = \sqrt{\mu \cdot 255 \cdot S}$, where V is the velocity of the car in kilometers per hour, μ is the **coefficient of friction,** and S is the length of the skid mark in meters. This formula assumes that the vehicle skidded to a complete stop. You can think of the coefficient of friction as the "holding ability" between the tire and the surface of the road. The higher the number, the better the tires "hold" onto the road.

Materials

- tape measure

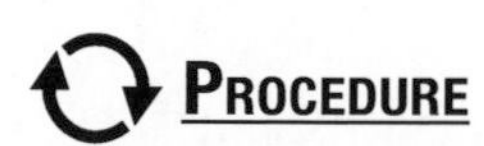

Procedure

1. Your teacher will provide you with several locations to measure. Use the tape measure to find the total distance that the "car" skidded. Record your numbers in the data table.
2. Make a note of the kind of surface over which the "car" was skidding. For example, was it wet, dry, icy, concrete, tar, grass, or gravel?

Extension Option

Find the maximum speed for several types of vehicles and calculate how far each could skid at its top speed.

Name ______________________ Date ______________________

7. Car Accident

STUDENT ACTIVITY PAGE

DATA COLLECTION AND ANALYSIS

Location measured	Skid length (in meters)	Surface qualities	Velocity of vehicle

For each skid mark you recorded above, calculate the speed of the vehicle in kilometers per hour. If the surface is wet or slippery, use a coefficient of friction of 0.2. If the surface is dry, use a coefficient of friction of 0.7. Remember, these values change depending on what surface the tire is passing over. The values 0.2 and 0.7 are just convenient examples to use for the activity.

CONCLUDING QUESTIONS

1. Calculate the speed of a vehicle by using a skid length that is double that of one you collected. Does a skid mark that is twice as long mean that the car was going twice as fast?

____

____

____

____

7. Car Accident

2. What problem would there be if you tried to use this formula for a vehicle that skidded and collided with another vehicle?

____

____

____

____

3. The two coefficients of friction provided are just assumed values. Considering the number of samples you collected, how many different coefficients of friction do you suppose you would need to make accurate estimations of vehicle speed?

____

____

____

____

Follow-Up Activity

Find a table of values for coefficient of friction. Determine the values that you should have used for coefficient of friction for the various surfaces you measured.

8. Tire Tracks

INSTRUCTIONAL OBJECTIVES

Students will be able to:

- collect tire tracks
- preserve evidence

NATIONAL SCIENCE EDUCATION STANDARDS CORRELATIONS

GRADES 5–8

Content standard	Bullet number	Content description	Bullet number(s)
A	1	Abilities necessary to do scientific inquiry	1–7
A	2	Understandings about scientific inquiry	1–5
B	1	Properties and changes of properties in matter	1
E	1	Abilities of technological design	1–5
E	2	Understandings about science and technology	4–5
G	1	Science as a human endeavor	1–2

GRADES 9–12

Content standard	Bullet number	Content description	Bullet number(s)
A	1	Abilities necessary to do scientific inquiry	1–5
A	2	Understandings about scientific inquiry	3, 5
E	1	Abilities of technological design	1–4
G	1	Science as a human endeavor	2

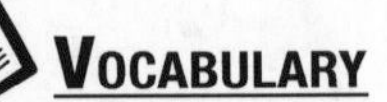

VOCABULARY

- **cast:** a three-dimensional impression made from a material such as plaster or dental casting
- **tire tracks:** distinctive impressions left behind in soft surfaces, such as mud, by the treads of tires

8. Tire Tracks

Teacher Resource Page

Materials

- bicycle (optional)
- car tire
- potting soil or topsoil
- plaster of Paris (fast-setting)
- hair spray
- tarp
- work gloves

Helpful Hints and Discussion

Time frame: one to two class periods
Structure: groups or partners
Location: classroom

This lab will have students make a tire track that they will preserve using a combination of techniques. Handling tires will require gloves, and the potting soil should be used either over a tarp or directly on the tarp. The quality of potting soils varies widely, and you will need to find a brand that will take a firm impression from a tire. If you have access to a garden or other outdoor area, you might want to collect some topsoil from that area to use instead of potting soil.

If you are having trouble getting a print from a car tire, use a tire with a small area of surface contact such as a bicycle tire or dirt-bike tire. A student can ride the bike though the soil to get a firm print.

Meeting the Needs of Diverse Learners

You might find that students with different abilities will benefit from extra help or extra challenges. Students who need extra challenges should complete the Extension Option and the Follow-Up Activity. The final observation of the cast requires careful attention to detail, and these students should help determine the method by which the cast will be studied. Students who need extra help might require additional instructions in handling the plaster of Paris and hair spray.

Scoring Rubric

Students meet the standard for this activity by:

- correctly following the procedure to collect tire casts
- using a proper strategy for planning the collection of tire casts
- correctly measuring tire-track details
- understanding the properties of the materials used for casting

8. Tire Tracks

TEACHER RESOURCE PAGE

RECOMMENDED INTERNET SITES

- **International Association for Property and Evidence—Footwear and Tire Track Evidence**
  www.iape.org/DownLoads/WCSO/footwear_and_tire_track_evidence.htm
- **Suite101.com—Forensic Impression Evidence: Matching Tire Tracks at a Crime Scene**
  http://traceevidence.suite101.com/article.cfm/forensic_impression_evidence

ANSWER KEY

Answers will vary. All answers depend on the quality of the cast that is taken from the soil. In general, some of the fine detail is lost in the casting process.

Name ______________________________ Date ______________________________

8. Tire Tracks

STUDENT ACTIVITY PAGE

Objective

To collect and preserve evidence from a crime scene

Before You Begin

A car accident is not the only reason why police might be interested in the tracks left behind by tires. **Tire tracks** are often found in materials such as mud, snow, and clay around crime scenes. Investigators can use these tire tracks to find vehicles that were in the vicinity of a crime. Then they can connect these vehicles to suspects or to help find witnesses.

When faced with tire tracks, investigators treat them in much the same way they do fingerprints. They take many pictures from a number of angles. They try to preserve the tracks as much as possible. One way to preserve the tracks is to cover them lightly with hair spray to keep them firm enough to take a **cast.** The casting material, such as plaster of Paris or dental casting material, has to solidify quickly. It must be applied carefully so it does not deform the tire print.

Once the cast and pictures are taken, they can be used to determine the kind of tire that made the impression. Tires can often be linked to specific makes, models, and even production years of cars. There are also massive databases that contain tracks of all the tires that are produced. These databases are maintained by the companies that make tires.

Once a general kind of tire is identified, investigators can then match small imperfections, deformities, nicks, and cuts to identify a particular tire.

Materials

- tire
- soil
- plaster of Paris
- hair spray
- tarp or pan
- work gloves

Procedure

1. Put on your gloves.
2. Place a pile of soil on a tarp or in a pan as instructed by your teacher. Be sure the soil is at least 10 centimeters deep. Also be sure that it will be long enough so that you can collect at least half of the tire's track.
3. Roll the tire through the soil while applying careful and steady pressure. This may require two people to be sure the print is full and deep.
4. Apply a light coating of hair spray to the track. Allow the hair spray to dry according to the instructions on the container.
5. Measure the length of the track.
6. Measure the width of the track.

Name ____________________ Date ____________________

8. Tire Tracks

STUDENT ACTIVITY PAGE

7. Find the width of an individual part of the tread if possible.
8. Measure the depth of the tread if possible.
9. Mix the plaster of Paris according to the instructions on the container. Pour the finished mixture into the tire track very carefully. Be sure to pour the mixture as close to the track as possible. If the mixture falls onto the track from too high, it will deform the print.
10. Allow the plaster of Paris to dry according to the instructions on the container.
11. When the plaster is dry, lift it out of the track and clean the cast carefully.

Extension Option

Make another track and collect it without using hair spray to firm up the track before you add the plaster of Paris. Compare the two tracks for quality.

Data Collection and Analysis

1. Length of track: __________
2. Width of track: __________
3. Width of individual tread component: __________
4. Depth of tread: __________

Concluding Questions

1. How does the quality of your cast compare to the original tire tread?

____

____

____

____

2. Which parts of the tread are very easy to see in your cast?

____

____

____

____

Name ____________________ Date ____________________

8. Tire Tracks

STUDENT ACTIVITY PAGE

3. Which parts of the tread were lost in the casting process?

4. Was any part of the tread destroyed completely by the casting process?

5. What details are apparent in your tread?

FOLLOW-UP ACTIVITY

Try and find some distinguishing characteristics on the cast and match them to marks on the original tire. See if you can actually align the correct part of the cast with the correct part of the tire.

9. Density of Glass Fragments

INSTRUCTIONAL OBJECTIVES

Students will be able to:

- calculate the density of glass fragments
- measure mass and volume
- compare the densities of different kinds of glass

NATIONAL SCIENCE EDUCATION STANDARDS CORRELATIONS

GRADES 5–8

Content standard	Bullet number	Content description	Bullet number(s)
A	1	Abilities necessary to do scientific inquiry	1–7
A	2	Understandings about scientific inquiry	1–5
B	1	Properties and changes of properties in matter	1
E	1	Abilities of technological design	1–5
E	2	Understandings about science and technology	4–5
G	1	Science as a human endeavor	1–2

GRADES 9–12

Content standard	Bullet number	Content description	Bullet number(s)
A	1	Abilities necessary to do scientific inquiry	1–5
A	2	Understandings about scientific inquiry	3, 5
B	2	Structure and properties of matter	4
E	1	Abilities of technological design	1–4
G	1	Science as a human endeavor	2

VOCABULARY

- **density:** the amount of mass per unit volume for a substance
- **refractive index:** a measure of the ability of a material to bend light that passes through it

9. Density of Glass Fragments

TEACHER RESOURCE PAGE

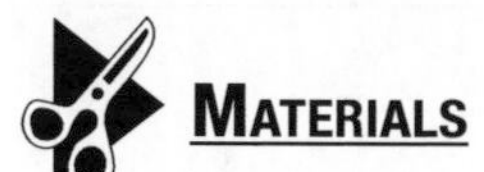

MATERIALS

- glass fragments (from three different kinds of glass)
- 250-ml beakers
- 100-ml graduated cylinders
- scales
- goggles
- apron
- work gloves

HELPFUL HINTS AND DISCUSSION

Time frame: one class period
Structure: individuals or partners
Location: classroom

Students will be finding the density of a collection of broken glass. Extreme caution should be used when handling the broken glass in all stages of the activity. Students should wear goggles, aprons, and work gloves if possible. Students will need three kinds of glass. Brown glass from beer bottles, clear glass from soda bottles, and broken laboratory glassware are all good choices. Be sure that at least one kind of glass is not similar to the other two. The glass needs to be broken into extremely small pieces. Be sure to wear goggles and a lab coat while prepping the glass for students. This can be achieved by placing the bottles in two or three plastic bags or a thick canvas bag, placing the bag in a plastic tub, and then smashing the bottle with a hammer. Be sure the glass is smashed quite small, and be careful to stop if the bag is cut open and glass spills out. The plastic tub is there to catch any fragments that might leak out. After the glass is smashed, pour it carefully into a 250-ml beaker and label it "Sample 1." Repeat with the other two kinds of glass and put them in beakers labeled "Sample 2" and "Sample 3." Fill a fourth beaker with one of the three kinds of glass and label it "Sample 4." Students will be trying to match Sample 4 to one of the other samples by comparing their densities. Because the density of glass varies, be sure you check to see that the 40 milliliters of water listed in the procedure is enough to cover the 10 grams of glass suggested in the procedure. If not, increase the original amount of water to a reasonable amount. Remind students to use extreme caution when handling the broken glass. You will need to provide an area where students can dispose of the broken glass. Provide at least one unbroken bottle of the same kind as one of the samples for the Extension Option.

MEETING THE NEEDS OF DIVERSE LEARNERS

You might find that students with different abilities will benefit from extra help or extra challenges. Students who need extra challenges should complete the Extension Option and the Follow-Up Activity. If possible, these students should do most of the direct handling of the broken glass. Remind students who need extra help that the process of determining density of the glass fragments has to follow a certain order. Students often discard samples or wet them without properly weighing them. Clarify the correct order of the procedure with these students.

9. Density of Glass Fragments

TEACHER RESOURCE PAGE

SCORING RUBRIC

Students meet the standard for this activity by:

- correctly following the procedure to determine the density of glass
- correctly recording data
- calculating final density based on reasonable data
- showing an understanding of the concept of density

RECOMMENDED INTERNET SITES

- **Forensic Science Communications—Glass Density Determination**
  www.fbi.gov/hq/lab/fsc/backissu/jan2005/standards/2005standards8.htm
- **Lenntech Glass**
  www.lenntech.com/glass.htm

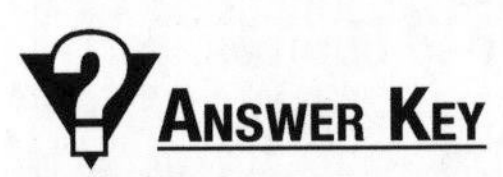

ANSWER KEY

Answers will vary depending on the kinds of glass used in the lab.

Name ______________________ Date ______________________

9. Density of Glass Fragments

STUDENT ACTIVITY PAGE

OBJECTIVE

To find and compare the densities of glass samples

BEFORE YOU BEGIN

There are thousands of kinds of glass. Glass is found in car windows, house windows, mirrors, car headlight covers, lightbulbs, beer bottles, soda bottles, wine bottles, food containers, cookware, and drinking glasses. It is also found in television screens, eyeglasses, bulletproof glass, welding masks, sunglasses, oven doors, picture frames, artwork, and car windshields. The list goes on and on. So the question is, how do you tell the kinds of glass apart, especially the many kinds of clear glass that can be found all around us? Another question is, why do we want to identify the kinds of glass?

Broken glass sticks to a wide variety of surfaces. A person committing a crime in which glass is broken could end up with glass on his or her hair, skin, clothing, and shoes. Glass can even stick into the soles of footwear. One way that criminals are linked to their crimes is by matching the **density** of broken glass found at the scene of a crime with the density of samples collected from suspects and their home or possessions. The FBI has a database with information about thousands of kinds of glass. This information includes density and **refractive index.** The refractive index is a measure of how light is bent as it passes through a piece of glass. Once the density or refractive index of a piece of glass has been identified, it narrows the range down within which investigators should be doing their testing. This allows them to match glass fragments from a suspect to a crime with great accuracy.

MATERIALS

- glass fragments (4 kinds)
- 250-ml beaker
- 100-ml graduated cylinders
- scales
- goggles
- apron
- work gloves

PROCEDURE

1. Put on your goggles, apron, and work gloves. **Be extremely careful while handling the glass fragments. Report any spilled glass or injuries to your teacher immediately.**
2. Weigh out about 10 grams of the glass labeled Sample 1.
3. Record the exact mass in the data table.
4. Fill your graduated cylinder to 40 milliliters exactly.
5. Add the approximately 10 grams of glass to the graduated cylinder and record the new volume.
6. Carefully pour off the water, being careful not to spill the glass, and take the glass to the disposal area set up by your teacher.

Name ______________________ Date ______________________

9. Density of Glass Fragments

STUDENT ACTIVITY PAGE

7. Repeat steps 2 through 6 for Samples 2, 3, and 4.
8. Calculate the density of each sample of glass fragments. Remember, the density formula is $D = m/V$, where D is density, m is mass, and V is volume. Record your findings in the data table.

EXTENSION OPTION

Determine the density of a bottle without smashing it. Find the mass of the bottle. Determine the volume by carefully submerging it in a large container that has been filled to the top. Catch any water that overflows in a tray and pour it into a graduated cylinder. The volume of the overflow will be the same as the volume of the bottle. Be sure to submerge the bottle in a manner such that it fills before water starts spilling over the sides of the container. Determine if it is glass from Sample 1, 2, 3, or 4.

DATA COLLECTION AND ANALYSIS

	Sample 1	Sample 2	Sample 3	Sample 4	Bottle (Extension Option)
Mass of glass					
Volume of water and glass					X
Volume of glass (water and glass – original 40 ml)					Volume overflowed (Volume of bottle)
Density (mass of glass/ volume of glass)					

CONCLUDING QUESTIONS

1. Which sample had the highest density?

____

____

____

____

Name ______________________ Date ______________________

9. Density of Glass Fragments

STUDENT ACTIVITY PAGE

2. Which sample had the lowest density?

____

____

____

____

3. Were any of the densities very close in value?

____

____

____

____

4. Which two samples were most likely the same kind of glass? Explain your reasoning.

____

____

____

____

FOLLOW-UP ACTIVITY

A more advanced method of determining the density of a material is called the density gradient column method. Research how this process works and determine how you would use it to find the density of a glass sample.

10. Glass Fracture Patterns

INSTRUCTIONAL OBJECTIVES

Students will be able to:

- analyze the patterns in broken glass
- identify the various kinds of fractures in glass
- use correct terminology to describe glass fracture patterns

NATIONAL SCIENCE EDUCATION STANDARDS CORRELATIONS

GRADES 5–8

Content standard	Bullet number	Content description	Bullet number(s)
A	1	Abilities necessary to do scientific inquiry	1–7
A	2	Understandings about scientific inquiry	1–5
B	2	Motions and forces	1, 3
E	1	Abilities of technological design	1–5
E	2	Understandings about science and technology	4–5
G	1	Science as a human endeavor	1–2

GRADES 9–12

Content standard	Bullet number	Content description	Bullet number(s)
A	1	Abilities necessary to do scientific inquiry	1–5
A	2	Understandings about scientific inquiry	3, 5
B	2	Structure and properties of matter	4
E	1	Abilities of technological design	1–4
G	1	Science as a human endeavor	2

VOCABULARY

- **concentric cracks:** fractures that form a roughly circular pattern around the hole or point of impact in glass
- **cone:** a funnel-shaped hole caused by a projectile moving through the glass at high speed
- **radial cracks:** fractures in glass that spread out from a central point, usually the point of entry for an object such as a bullet

10. Glass Fracture Patterns

- **ream:** an imperfection in the glass that might have been formed at the time the glass was made, or might have been caused when the glass was exposed to stress and then did not return to its original shape
- **Wallner lines:** lines found in a sample of chipped glass that look like the rings found in the cross section of a tree

Materials

- goggles
- work gloves
- glass plates (6 inches × 6 inches)
- cutting board
- hammer
- magnifying glass
- thick plastic bag or garbage bag

Helpful Hints and Discussion

Time frame: one class period
Structure: individuals or partners
Location: classroom

Students will be breaking glass, so there are a number of safety precautions that should be made. **Students should wear goggles and gloves and will need a place to dispose of their broken glass.** The thick plastic bag mentioned in the procedure should drape over the glass and cutting board to completely prevent glass from spraying during the breaking process. You might want to provide students with different kinds of glass objects to break if time permits.

The quality of glass samples varies widely, and you might not get good pattern results when breaking the glass with a hammer. Covering the glass with clear tape, such as packing tape, might better preserve the patterns. This also adds another layer of safety.

Meeting the Needs of Diverse Learners

You might find that students with different abilities will benefit from extra help or extra challenges. Students who need extra challenges should complete the Extension Option and the Follow-Up Activity. These students can be allowed to test extra samples of glass with other tools. You might also allow them to preserve a broken glass sample with tape if time allows. This lab has a variety of safety issues, and students who need extra help should be walked through the procedure. You should highlight places where there are safety issues they should be aware of.

10. Glass Fracture Patterns

TEACHER RESOURCE PAGE

SCORING RUBRIC

Students meet the standard for this activity by:

- safely following the procedure for breaking glass
- recording data in the form of detailed sketches
- drawing clear conclusions about the breakage patterns
- understanding the properties of glass and how it breaks

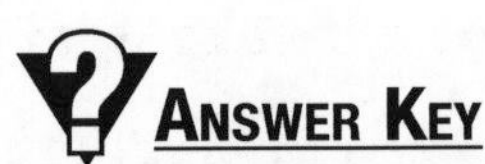

RECOMMENDED INTERNET SITES

- **Federal Bureau of Investigation—Forensic Science Communications: Glass Fractures**
  www.fbi.gov/hq/lab/fsc/backissu/jan2005/standards/2005standards7.htm
- **Glass on Web—Glass manual**
  www.glassonweb.com/glassmanual/

ANSWER KEY

1. The pieces at the center of the fractures are generally much smaller with the hard hammer blow. The farther away a fractured piece is from the point of impact, the less it is reduced in size.
2. Students are likely to see radial cracks and Wallner lines. It is possible that they will see concentric cracks or cones, but it is not as likely.
3. The broken surface of a fracture will appear irregular and crooked when seen with a magnifying glass. This indicates the relatively random nature of glass at the molecular level.
4. Students might not notice a large difference when they look at the pattern formed by the 45° hammer blow, but in some instances they will notice that the fracture lines are closer together on one side of the impact site than the other.

Name ______________________________ Date ______________________________

10. Glass Fracture Patterns

STUDENT ACTIVITY PAGE

OBJECTIVE

To understand the various patterns found in damaged or broken glass

BEFORE YOU BEGIN

Glass can be broken in a variety of ways. It can be completely destroyed to powder form. It can be broken into a variety of sizes from sand-grain pieces to fist-sized chunks. Or it can simply crack and retain its overall shape. How glass breaks can tell investigators a lot about how it was broken, and often it can reveal the object that broke it. Did the window break from the inside or from the outside? Was the object that broke the glass moving quickly or slowly?

There are some terms that are used to describe the various patterns found in a piece of broken glass. There are **concentric cracks,** which are fractures that form a roughly circular pattern around the hole or point of impact. These fractures usually end where they run into another crack. This tells investigators which bullet hole was made first when many holes are in the same window.

A **cone** is a funnel-shaped hole caused by a projectile moving through the glass at high speed. **Radial cracks** are fractures that spread out from a central point, usually the point of entry for an object such as a bullet. A **ream** is an imperfection in the glass that might have been formed at the time the glass was made, or it might have been caused when the glass was exposed to stress and then did not return to its original shape. **Wallner lines** are lines that look like the rings one might find when looking at the cross section of a tree. These lines don't indicate age as with a tree, however. Instead they suggest the direction that a crack was moving through a piece of glass.

MATERIALS

- goggles
- work gloves
- glass plates (6 inches × 6 inches)
- cutting board
- hammer
- magnifying glass
- thick plastic bag or garbage bag

PROCEDURE

1. Put on your goggles and gloves.
2. Set a glass plate on the cutting board.
3. Cover the plate with the thick plastic bag. **Be sure the glass is well covered on all sides with the bag before proceeding.**
4. Swing the hammer straight down and hit the bag lightly over the center of the glass. **Do not smash the glass as hard as you can with the hammer.** Check to see if the glass broke. If the glass did not break, hit it slightly harder until it does. Try to hit it directly in the center. **After it breaks, do not hit it again.**

Name ______________________ Date ______________________

10. Glass Fracture Patterns

STUDENT ACTIVITY PAGE

5. Record any of the patterns mentioned in the activity and sketch the results in your data table. Use a magnifying glass to inspect the edges of the fractures.
6. Discard the broken glass as directed by your teacher.
7. Repeat the breaking process, but this time, hit the glass very hard. **Be sure the glass is well covered on all sides with the bag before proceeding.**
8. Repeat the breaking process with another piece of glass, but this time, hit the glass at a 45° angle. **Be sure the glass is well covered on all sides with the bag before proceeding.**

EXTENSION OPTION

Break another piece of glass using the same method as in steps 1 through 5, but instead of using a hammer, try breaking the glass with some other tools, such as the sharp end of a screwdriver or the heel of a shoe. Compare your results to the breakage pattern seen with the hammer.

DATA COLLECTION AND ANALYSIS

Breaking method	Patterns	Sketch
Light hammer blow		
Hard hammer blow		

(chart continues on next page)

Name ____________________ Date ____________________

10. Glass Fracture Patterns

STUDENT ACTIVITY PAGE

Breaking method	Patterns	Sketch
Angled hammer blow		

Concluding Questions

1. How did the size of the glass fragments from the light hammer blow compare to the size of the fragments from the hard hammer blow? Was this true of all the pieces in general?

2. What structures mentioned in the activity did you see in your broken glass?

3. Describe the edges of the broken glass that you viewed with the magnifying glass.

10. Glass Fracture Patterns

4. How did striking the glass at an angle affect the shape of the fracture pattern?

 ____

 ____

 ____

 ____

 **Follow-Up Activity**

There are a wide variety of glass types. Research to find a rough estimate of the number of different kinds of glass that are produced by modern production means. What are some of the more unusual ways that glass is used?

11. Physical Properties of Soil

TEACHER RESOURCE PAGE

Instructional Objectives

Students will be able to:

- analyze soil appearance
- use a microscope to analyze soil appearance
- identify the various components of soil

National Science Education Standards Correlations

Grades 5–8

Content standard	Bullet number	Content description	Bullet number(s)
A	1	Abilities necessary to do scientific inquiry	1–7
A	2	Understandings about scientific inquiry	1–5
B	1	Properties and changes of properties in matter	1
D	1	Structure of the earth system	4–5
E	1	Abilities of technological design	1–5
E	2	Understandings about science and technology	4–5
G	1	Science as a human endeavor	1–2

Grades 9–12

Content standard	Bullet number	Content description	Bullet number(s)
A	1	Abilities necessary to do scientific inquiry	1–5
A	2	Understandings about scientific inquiry	3, 5
B	2	Structure and properties of matter	4
D	2	Geochemical cycles	1–2
E	1	Abilities of technological design	1–4
G	1	Science as a human endeavor	2

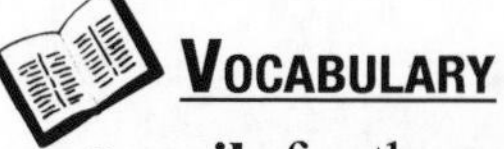

Vocabulary

- **soil:** for the purposes of forensics, any of the various materials found at the surface of the earth, including dirt, rocks, biological materials, and synthetic materials

11. Physical Properties of Soil

TEACHER RESOURCE PAGE

MATERIALS

- gloves (latex and nitrile)
- 250-ml beaker
- scales
- soil sample
- microscope
- magnet
- tweezers
- color chart (found at most paint-supply stores)
- paper cups or petri dishes
- tray

HELPFUL HINTS AND DISCUSSION

Time frame: one class period
Structure: individuals or partners
Location: classroom

You will need to supply students with a sample of soil to use for this activity. The sample should contain plant and animal matter and should have sand, pebbles, and silt if possible. The soil should be "salted" with some forensic evidence such as pieces of used gum, broken glass (very small), cigarette butts (simulated), bottle caps, buttons, and so forth. These materials should be things that either might have DNA or fingerprints on them, or could be traced back to a manufacturer.

Some students might be allergic to latex gloves, so be sure to have nitrile gloves available.

MEETING THE NEEDS OF DIVERSE LEARNERS

You might find that students with different abilities will benefit from extra help or extra challenges. Students who need extra challenges should complete the Extension Option and the Follow-Up Activity. These students might also be encouraged to use a reference guide to try and identify some of the rock types found in the soil sample. Students who need extra help might need assistance distinguishing the various components of the soil, such as identifying animal or plant parts. They might benefit from a partner or from clarification as to what they should be collecting from the sample.

SCORING RUBRIC

Students meet the standard for this activity by:

- carefully separating materials from the soil
- following a strategy for clear observation of soil
- clearly recording data
- completing activity with an understanding of the various components that can be found in soil

11. Physical Properties of Soil

TEACHER RESOURCE PAGE

RECOMMENDED INTERNET SITES

- **Geoforensics—Forensic Geology**
  www.geoforensics.com/geoforensics/fgeology.html
- **NASA—Soil Science Education Home Page: Secrets Hidden in Soil**
  http://soil.gsfc.nasa.gov/forengeo/secret.htm

ANSWER KEY

Answers will vary depending on the nature of the soil sample, but students should be encouraged to answer with as much detail as possible.

Name ______________________________ Date ______________________________

11. Physical Properties of Soil

Student Activity Page

Objective

To determine the various characteristics of a soil sample

Before You Begin

If you leave your home and pick up the first **soil** sample you see, odds are that it will be very different from soil that is found a kilometer away. Soil is a material that has a very specific "fingerprint." The materials that are found in soil have many sources. Soil is made up of naturally occurring components such as various minerals and kinds of rocks, sand, and gravel. There is also a large collection of biological agents in most samples of soil. There can be material from plants, animals, and insects and the waste products or dead bodies of all three. You will probably find seed pods and pollen. These materials survive for a long time in soil. There can also be a lot of synthetic materials that have been put there by human activity. Soil often contains traces of oil, gasoline, paint, food, gum, and cigarette butts. Soil can contain almost anything a person can drop on the ground.

There are many physical properties that you can use to describe and identify soil. These include but are not limited to color, density, hardness, magnetic nature, odor, and shape of particles. Very basic lab equipment can be used to find these physical properties. A magnet passed over a soil sample will show if there is any magnetic material in the soil. Although smelling things in the lab is generally unsafe, sometimes investigators can't help but notice the smell of materials such as gasoline, ammonia, and gunpowder.

Materials

- gloves
- 250-ml beaker
- scales
- soil sample
- microscope
- magnet
- tweezers
- color chart
- paper cups or petri dishes
- tray

Procedure

1. Put on your gloves.
2. Weigh and record the mass of a 250-ml beaker.
3. Fill the 250-ml beaker to the 250-ml line with the soil sample provided by your teacher.
4. Weigh and record the mass of the soil and beaker together.
5. Pour the soil into the tray and spread it out as thin as possible.
6. Use the tweezers to collect any evidence of animal life and place it into a separate cup. This might include dead insects, hair, or still-living animals. List any items found in the data table.

Name ______________________________ Date ______________________

11. Physical Properties of Soil

STUDENT ACTIVITY PAGE

7. Use the tweezers to collect any evidence of plant life and place it into a separate cup. This might include twigs, grass, seeds, and so forth. List any item found in the data table.
8. Use the tweezers to collect any evidence of human-made materials and place them into a separate cup. This might include bottle caps, cigarette butts, gum, paper, plastic, and so forth. List any item found in the data table.
9. Get a color chart and find the color that best matches the color of your soil sample. Record the color match in your data table.
10. Use the microscope to look at some of the smallest particles of the soil. Describe what you see in the data table.
11. Use the microscope to look at some of the largest particles of the soil. Describe what you see in the data table.
12. Drag the magnet through the soil. Make a note in the data table if any material sticks to the magnet.

EXTENSION OPTION

Think of some other physical properties that you could use to describe your soil sample and determine them.

DATA COLLECTION AND ANALYSIS

Mass of beaker	
Mass of beaker and soil	
Rough density estimate for soil $D = m/V =$ (mass of beaker and soil – mass of beaker)/(250 ml)	
Evidence of animal life	
Evidence of plant life	

Name ____________________ Date ____________________

11. Physical Properties of Soil

Student Activity Page

Evidence of human-made materials	
Color match of soil	
Description of small soil particles	
Description of large soil particles	
Is there evidence of magnetic material in soil?	(Circle one) Yes No

Concluding Questions

1. Which parts of your soil sample do you think might have been unique to the location where the sample was collected? Why do you think so?

____

____

____

____

2. Did your soil seem as though it came from a specific location such as a beach or garden? Why do you think so?

____

____

____

____

11. Physical Properties of Soil

3. Where there any human-made materials in your sample that could be traced to an individual or to a manufacturer? Why do you think so?

____

____

____

____

4. Did you notice any odor coming from your sample? If so, what?

____

____

____

____

5. Estimate how many different components your soil sample contained that you could see with the unaided eye.

____

____

____

____

FOLLOW-UP ACTIVITY

Soil is defined a number of different ways by people in different professions. Find as many definitions of soil as possible and determine whether or not each definition is useful to the forensic scientist or not.

12. Components of Soil

TEACHER RESOURCE PAGE

Instructional Objectives

Students will be able to:

- separate soil samples into components of different sizes
- identify components of different sizes in a soil sample
- calculate percentages
- use a sieve to separate soil components
- use a microscope to analyze the small components of soil

National Science Education Standards Correlations

Grades 5–8

Content standard	Bullet number	Content description	Bullet number(s)
A	1	Abilities necessary to do scientific inquiry	1–7
A	2	Understandings about scientific inquiry	1–5
B	1	Properties and changes of properties in matter	1
D	1	Structure of the earth system	4–5
E	1	Abilities of technological design	1–5
E	2	Understandings about science and technology	4–5
G	1	Science as a human endeavor	1–2

Grades 9–12

Content standard	Bullet number	Content description	Bullet number(s)
A	1	Abilities necessary to do scientific inquiry	1–5
A	2	Understandings about scientific inquiry	3, 5
B	2	Structure and properties of matter	4
D	2	Geochemical cycles	1–2
E	1	Abilities of technological design	1–4
G	1	Science as a human endeavor	2

Vocabulary

- **clay:** rocks with a diameter between 0.001 and 0.002 millimeters
- **colloids:** rocks with a diameter less than 0.001 millimeters

12. Components of Soil

Teacher Resource Page

- **pebbles:** rocks with a diameter between 2 and 75 millimeters
- **sand:** rocks with a diameter between 0.05 and 2 millimeters
- **silt:** rocks with a diameter between 0.002 and 0.05 millimeters

Materials

- goggles
- gloves (latex and nitrile)
- dust mask
- soil sieves (three sizes—pebble, sand, and silt)
- soil sample (dry)
- scales
- magnifying glass
- microscope

Helpful Hints and Discussion

Time frame: one class period
Structure: individuals or groups
Location: classroom or outside

Separating soil samples can produce dust that is unpleasant or unsafe to breathe. You and your students should wear dust masks when performing this lab. Performing the separation outside would be preferable but is not required. Some students might be allergic to latex gloves, so be sure to have nitrile gloves available.

The soil sample produced should be dry and should be collected from a source that contains gravel, sand, and silt. Such samples can often be found along the shore near rivers, streams, ponds, or lakes. A second similar sample should be provided for the Extension Option.

Meeting the Needs of Diverse Learners

You might find that students with different abilities will benefit from extra help or extra challenges. Students who need extra challenges should complete the Extension Option and the Follow-Up Activity. These students should have access to a soil texture triangle and should be able to identify the classification of their sample without assistance. Students who need extra help should be instructed on proper weighing techniques and the necessity of careful sample handling. They should also be assisted in calculating the percentages.

12. Components of Soil

TEACHER RESOURCE PAGE

SCORING RUBRIC

Students meet the standard for this activity by:

- properly using equipment to separate soil
- correctly calculating percentages for soil components
- properly collecting and recording data
- using the correct terminology when describing components of soil

RECOMMENDED INTERNET SITES

- **University of Florida IFAS Extension—Soil Texture**
  http://edis.ifas.ufl.edu/SS169
- **The University of Georgia Savannah River Ecology Laboratory—Soils and Particle Size**
  www.uga.edu/srel/kidsdoscience/soils-planets/soil-particle-size.pdf

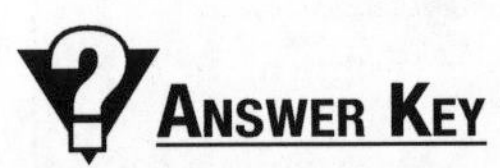

ANSWER KEY

1–3. Answers will vary depending on the quality of the soil sample.

4. Soil samples collected from different locations should have different percentages of the various components.
5. Percentages are a quick method investigators can use to match soil samples. If the percentages are very different, they may suspend further testing. If the percentages are close, they may choose to match microscopic evidence in the soil that contains more specific information.

Name ______________________________ Date ______________________________

12. Components of Soil

Student Activity Page

Objective

To analyze a soil sample by comparing relative amounts of components of different sizes

Before You Begin

Soil has a large number of components. When you are analyzing a soil sample, however, it is usually far simpler just to look at the relative amounts of particles that occur in the soil that are different sizes. A number of scales are used to describe the particles found in a sample of dirt or soil. But there are some basic classifications that are generally agreed upon. The largest particles are **pebbles,** sometimes called gravel. Pebbles come in a wide variety of shapes and sizes. Generally a single pebble is something you could pick up easily with your fingers. The next smaller size is **sand.** Individual grains of sand can usually be seen with the unaided eye. They feel gritty when rubbed between your fingers.

Silt, clay, and **colloids** are extremely small. They are difficult to pick out as individual pieces when you are looking at a mixed sample of soil. These smaller particles feel smooth when rubbed between your fingers, as if you were rubbing baby powder or cooking flour. The amount of each of these that is found in a soil sample is fairly consistent and can be used to quickly match a sample to a location. The sample collected at a crime scene can be separated by size. The amount of each size is recorded as a percentage. Samples taken later at other places can then be compared to the first sample. If the percentages are a rough match, more detailed testing can then be done.

Classification	Diameter
colloid	less than 0.001 mm
clay	0.001 mm to 0.002 mm
silt	0.002 mm to 0.05 mm
sand	0.05 mm to 2.00 mm
fine pebbles	2.00 mm to 5.00 mm
medium pebbles	5.00 mm to 20.00 mm
coarse pebbles	20.00 mm to 75.00 mm

Materials

- goggles
- gloves
- dust mask
- nested sieves
- soil sample
- scale
- magnifying glass
- microscope

Name ________________________________ Date ____________________

12. Components of Soil

STUDENT ACTIVITY PAGE

PROCEDURE

1. Put on your goggles, gloves, and dust mask.
2. Collect a soil sample that has been prepared by your teacher.
3. Weigh your sample and record the mass in the data table.
4. Use the sieves to separate your sample. Place the sample in the top tray and shake lightly until the small particles stop falling through.
5. Remove the top tray (tray 1) and weigh the particles that are in the top tray. Look at these particles with the magnifying glass and note anything they have in common. These particles are the pebbles (gravel). Record the mass in the data table.
6. Continue shaking the sieve set until particles stop falling through the top tray. Remove the top tray (which is now tray 2) and weigh the particles that are in the top tray. These particles are the sand. Look at these particles with the magnifying glass and note anything they have in common. Record the mass in the data table.
7. The bottom tray (tray 3) should now be full of the materials that were not caught by the top two trays. Weigh the particles from the tray. These particles are the silt, clay, and colloids. Look at these particles with the microscope and note anything they have in common. Record the mass in the data table.
8. Return or discard your soil as directed by your teacher.

EXTENSION OPTION

Repeat the procedure with a second soil sample provided by your teacher that was collected from a different location.

DATA COLLECTION AND ANALYSIS

Total mass of soil	
Mass of pebbles (gravel)	
Common characteristics of pebbles	

(chart continues on next page)

Name ______________________ Date ______________________

12. Components of Soil

STUDENT ACTIVITY PAGE

Mass of sand	
Common characteristics of sand	
Mass of silt, clay, and colloids	
Common characteristics of silt, clay, and colloids	
Determine the percentage of each component that was in the original soil. Take the mass of pebbles and divide by the total mass and multiply by 100%. This is the percentage of the original sample that was pebbles.	
Repeat the percentage calculation for sand.	
Repeat the percentage calculation for silt, clay, and colloids.	

Concluding Questions

1. What component made up the largest amount of your sample by mass?

____

____

Name ______________________ Date ______________________

12. Components of Soil

STUDENT ACTIVITY PAGE

2. What component made up the smallest amount of your sample?

____

____

3. What component looked like it had the most mass before you started?

____

____

4. How would samples taken from a beach or garden be different from the sample you tested?

____

____

____

____

____

5. How could investigators use the different percentages to match soil samples?

____

____

____

____

____

Follow-Up Activity

Find a soil texture triangle online or in a reference book and determine the exact classification of the soil sample you processed.

13. Handwriting Analysis

Teacher Resource Page

Instructional Objectives

Students will be able to:

- identify characteristics of handwriting
- use correct terminology to describe components of handwriting
- compare handwriting samples
- describe pattern similarities between two samples of handwriting

National Science Education Standards Correlations

Grades 5–8

Content standard	Bullet number	Content description	Bullet number(s)
A	1	Abilities necessary to do scientific inquiry	1–7
A	2	Understandings about scientific inquiry	1–5
E	1	Abilities of technological design	1–5
E	2	Understandings about science and technology	4–5
G	1	Science as a human endeavor	1–2

Grades 9–12

Content standard	Bullet number	Content description	Bullet number(s)
A	1	Abilities necessary to do scientific inquiry	1–5
A	2	Understandings about scientific inquiry	3, 5
E	1	Abilities of technological design	1–4
G	1	Science as a human endeavor	2

Vocabulary

- **arrangement:** the way people connect letters when writing in cursive
- **content** and **word usage:** the internal meaning of a passage
- **letter shape:** determined by finding multiple examples of the same letter and then comparing the qualities that make them unique
- **pressure:** force per unit area
- **slant:** the angle with which a person forms letters relative to the lines of the paper or to the edges of the paper if it is unlined

13. Handwriting Analysis

MATERIALS

- paper
- writing utensils
- magnifying glasses
- scissors

HELPFUL HINTS AND DISCUSSION

Time frame: one class period
Structure: individuals
Location: classroom

This activity is easy to start but requires good organizational skills on the part of the instructor. You must collect student handwriting samples and then redistribute them in a manner that guarantees that each individual has a matching set of sentences. You may choose to give students a set with no matches, but it prevents them from finding the matching qualities that are so important to handwriting investigators. You will want to assign each student a code ahead of time to write on their samples so that you can group them accordingly once they hand them in. It will help if you have beakers or cups to put the samples in so that you can separate them quickly according to a list that you can make ahead of time that insures a match in each group. Each student should get six pieces of paper back if it has been done correctly—five different versions of Sentence 1 and one version of Sentence 2 that matches one of the five versions of Sentence 1. Read the Procedure section in the Student Activity Pages for a version of how the collection and redistribution of sentences looks from the student side. This might make the process easier for both you and your students.

MEETING THE NEEDS OF DIVERSE LEARNERS

You might find that students with different abilities will benefit from extra help or extra challenges. Students who need extra challenges should complete the Extension Option and the Follow-Up Activity. These students should be encouraged to use as much of the technical language as possible. They might want to try and disguise their handwriting to make the activity more challenging. Students who need extra help might have trouble finding individual structures that match. Tell them to focus on the letters one at a time, starting with just *a,* then just *b,* and so forth until they match or exclude a sentence.

SCORING RUBRIC

Students meet the standard for this activity by:

- forming a strategy for comparing handwriting samples
- matching samples by collecting data
- using correct terminology to describe handwriting samples
- using a magnifying glass to enhance details

13. Handwriting Analysis

RECOMMENDED INTERNET SITES

- **Forensic-Evidence.com—Graphology, Graphoanalysis, and Handwriting Analysis**
  http://forensic-evidence.com/site/ID/ID00004_3.html
- **HowStuffWorks—How Handwriting Analysis Works**
  http://science.howstuffworks.com/handwriting-analysis1.htm

ANSWER KEY

1–2. Answers will vary relative to the handwriting samples.

3. Unless a student has tried to disguise his or her handwriting, students can often match the samples before they start looking at the individual qualities.
4. Answers will vary relative to the handwriting samples.
5. Students can generally tell with two samples, but the more samples they have, the greater the likelihood they will uncover an attempt to disguise handwriting.
6. Suspects who are asked to provide a writing sample on demand may choose to disguise their handwriting. A sample taken from preexisting documents is more likely to provide an accurate sample.

Name ______________________________ Date ______________________

13. Handwriting Analysis

Student Activity Page

Objective

To analyze handwriting and to understand the various qualities that handwriting contains

Before You Begin

When you are first learning to write, you develop habits that will probably stay with you your whole life. For instance, do you put a horizontal line through the middle of your sevens? Do you often use the & symbol instead of writing the word *and*? How do you dot your *i*'s? There are a number of details that are very similar from item to item when you write. Investigators compare handwritten documents by the following basic qualities:

- **Letter shape:** This is determined by finding multiple examples of the same letter and then comparing the qualities that make the letters unique.
- **Pressure:** The amount of pressure changes how deeply the paper is dented and can increase the width of lines.
- **Arrangement:** When people write in cursive, the way they connect letters is somewhat unique. Even if people do not use cursive frequently, they probably use personal shortcuts that make their writing distinctive.
- **Content and word usage:** This refers to the actual meaning of the writing sample. People use certain phrases, and the usage can stand out. The Unabomber, Ted Kaczynski, was known to use the phrase "you can't eat your cake and have it, too" instead of the more common "you can't have your cake and eat it, too." This oddity of word usage was actually recognized by Kaczynski's brother and led to his eventual arrest.
- **Slant:** This is the angle with which a person forms letters relative to the lines of the paper or to the edges of the paper if it is unlined. Some investigators also believe that heavily slanted writing shows stress or other problems on the part of the writer.

Materials

- paper
- writing utensil
- magnifying glass
- scissors

Procedure

1. In your best cursive handwriting, copy Sentence 1 five times. Use the Data Collection and Analysis section or a separate sheet of paper. (If you cannot write in cursive, simply write the sentence as neatly as you can.) There should be enough space between the sentences so that the sentences can be separated with scissors without cutting off any of the letters. Each sentence uses all 26 letters from the alphabet at least once.

   Sentence 1: The public was amazed to view the quickness and dexterity of the juggler.

Name ______________________ Date ______________________

13. Handwriting Analysis

STUDENT ACTIVITY PAGE

2. Copy Sentence 2 one time. At the end of the sentence, write your initials. **Only write your initials at the end of Sentence 2.**

   Sentence 2: The job requires extra pluck and zeal from every young wage earner.
3. Carefully separate your five versions of Sentence 1 with scissors.
4. Write the code number on your samples as instructed by your teacher. Give the samples to your teacher.
5. Give your copy of Sentence 2 to your teacher.
6. After your teacher organizes the samples, collect samples of Sentence 1 and Sentence 2 that were written by other students.
7. Find the version of Sentence 1 and Sentence 2 that match.
8. Use the magnifying glass to find qualities such as slant, letter shape, and arrangement that "prove" the two samples came from the same person. Circle corresponding qualities in the two sentences and label them with numbers that show they match.

EXTENSION OPTION

Find the person who wrote the sentences you matched. Have them write "The quick brown fox jumps over a lazy dog." Find another component of their handwriting that matches.

Name ______________________ Date ______________________

13. Handwriting Analysis

STUDENT ACTIVITY PAGE

DATA COLLECTION AND ANALYSIS

Sentence 1: ______________________

Sentence 1: ______________________

Sentence 1: ______________________

Sentence 1: ______________________

Sentence 1: ______________________

Sentence 2: ______________________

Name ______________________ Date ______________________

13. Handwriting Analysis

STUDENT ACTIVITY PAGE

CONCLUDING QUESTIONS

1. What letter or letters were obviously written by the same person? What were the similarities?

2. What letter or letters showed very little similarity? How were they different?

3. Were you able to match the sentences before you checked any individual qualities? Why or why not?

4. What qualities did you use to "prove" the two samples were the same?

5. Were two handwriting samples enough for you to be absolutely sure they were from the same person? Why or why not?

Name ________________________________ Date ________________________________

13. Handwriting Analysis

STUDENT ACTIVITY PAGE

6. What drawback do investigators face when they ask a suspect to provide a writing sample?

____

____

____

____

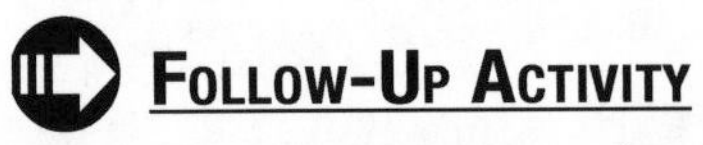

There are many groups that study the supposed underlying psychological makeup of a person based on his or her handwriting. Research some of the characteristics that handwriting can supposedly show.

14. Searching Through Garbage

TEACHER RESOURCE PAGE

Instructional Objectives

Students will be able to:

- draw conclusions about materials collected as evidence
- reconstruct damaged evidence
- identify useful evidence discarded by criminals

National Science Education Standards Correlations

Grades 5–8

Content standard	Bullet number	Content description	Bullet number(s)
A	1	Abilities necessary to do scientific inquiry	1–7
A	2	Understandings about scientific inquiry	1–5
E	1	Abilities of technological design	1–5
E	2	Understandings about science and technology	4–5
G	1	Science as a human endeavor	1–2

Grades 9–12

Content standard	Bullet number	Content description	Bullet number(s)
A	1	Abilities necessary to do scientific inquiry	1–5
A	2	Understandings about scientific inquiry	3, 5
E	1	Abilities of technological design	1–4
G	1	Science as a human endeavor	2

Vocabulary

- **garbage:** in a legal sense, materials abandoned by a suspect in a manner such that the suspect cannot claim legal ownership

Materials

- gloves (latex and nitrile)
- garbage
- garbage bags
- tweezers
- tray

14. Searching Through Garbage

TEACHER RESOURCE PAGE

HELPFUL HINTS AND DISCUSSION

Time frame: one class period
Structure: individuals or partners
Location: classroom

Students will try to find evidence of a crime in some garbage that has been "collected" from the curb outside of a suspect's house. You will need to create a realistic yet safe bag of garbage for students to search through. The garbage sample should include receipts from stores that seem to list items that could be used to commit a crime. Such receipts can be made on a computer and printed out. You can even have students produce artificial receipts that will be found by other students in their samples. Receipts or notes could be torn up so that students have to put them back together. Photographs of weapons or crime scenes can also be included. Other materials can be added at your discretion. Items that could have fingerprints or DNA and printouts of Web sites that describe how to commit crimes could be included. You might choose to make a bag that represents a different crime for each person or group. Regular household garbage, such as empty cereal boxes, used paper cups, and so forth, can add realism and provide a challenge as students try to distinguish which garbage is evidence and which garbage is garbage. **Be sure that no actual garbage that could pose a health hazard, such as used tissues or broken glass, is included in the sample.** Looking at the other activities in this book might give you more ideas about items to include.

Some students might be allergic to latex gloves, so be sure to have nitrile gloves available.

MEETING THE NEEDS OF DIVERSE LEARNERS

You might find that students with different abilities will benefit from extra help or extra challenges. Students who need extra challenges should complete the Extension Option and the Follow-Up Activity. These students should be partnered with struggling students to help them produce a more methodical search of the collected evidence. Students who need extra help might not assume there is evidence such as DNA or fingerprints if they cannot see it. Be sure to remind them of the various things they might find in their collected evidence.

SCORING RUBRIC

Students meet the standard for this activity by:

- carefully separating evidence
- using a strategy to evaluate evidence
- clearly recording data
- forming a conclusion that is consistent with the collected evidence

14. Searching Through Garbage

RECOMMENDED INTERNET SITES

- **LegalMatch—Police Search Lawyers**
  www.legalmatch.com/law-library/article/can-police-officers-search-my-garbage.html
- **Stephen A. Houze, Criminal Defense Attorney—Searching Garbage Requires Warrant, Appeals Court Rules**
  www.shouze.com/display-cases.asp?artID=15

ANSWER KEY

1. Answers will vary but should make reference to receipts or photographs if they were used. Students should also realize that some of the garbage might have fingerprints or DNA on it.
2. Items such as paper cups or cotton swabs might contain DNA evidence. Smooth surfaces, such as on a plastic cup, might contain fingerprints.
3. Answers will vary depending on the garbage sample.
4. Garbage bags are very likely to have fingerprints on them.

Name ______________________________ Date ______________________________

14. Searching Through Garbage

STUDENT ACTIVITY PAGE

OBJECTIVE

To determine how garbage can be used as valuable evidence in a criminal case

BEFORE YOU BEGIN

Movies and television can make the life of a crime-scene investigator seem very exciting. But it is not unusual for a person in that job to spend some time digging through **garbage.** There have been many cases in which suspects have left incriminating evidence in their garbage. Sometimes suspects are surprised when evidence is found by police officers and used against them.

There is some debate as to whether or not police officers need a warrant to search garbage that has been left out to be picked up by a company that collects trash. The U.S. Supreme Court has even ruled on the issue. They decided it was not unconstitutional for police to search a person's garbage. While the practice is used in most states, some state courts have disagreed with the Supreme Court's ruling. For example, in some states, the police cannot search garbage that is in a closed container that belongs to the homeowner.

Evidence of a crime might be very obvious, such as a bloody knife, dead body, or recently fired gun. But it could be less obvious, such as a receipt, DNA, a bank statement, or trace evidence of narcotics. Police have to be sure that they have actually collected the garbage of the person they are investigating. It is often best if officers actually see the suspect leaving the garbage. Garbage left in a common area for pickup is harder to match to a suspect. Its inclusion as evidence could be challenged later at trial.

MATERIALS

- gloves
- garbage
- garbage bags
- tweezers
- tray

PROCEDURE

1. Put on your gloves to protect both you and the evidence.
2. Get a bag of garbage from your teacher.
3. Carefully pour the garbage into the tray and spread it out. If you have more garbage than will comfortably fit into your tray, only use a little at a time.
4. Use your gloved hands and tweezers if necessary to sort and manipulate the garbage. **Be aware the garbage can contain dangerous materials such as used tissues, broken glass, or needles. Handle the samples as if they are all health hazards.**
5. List and describe each piece of garbage that you remove. Write down if it has any distinguishing characteristics that indicate it might be evidence of a crime.
6. Reconstruct any damaged garbage.
7. Return the garbage to its bag and return it to your teacher.

Name ______________________________ Date ______________________

14. Searching Through Garbage

STUDENT ACTIVITY PAGE

Extension Option

Make a list of how every piece of garbage collected might be used as evidence.

Data Collection and Analysis

Item	Description	Likely evidence? (Yes/No)
1.		
2.		
3.		
4.		
5.		
6.		
7.		
8.		
9.		
10.		
11.		

(chart continues on next page)

Name ______________________________ Date ______________________

14. Searching Through Garbage

STUDENT ACTIVITY PAGE

Item	Description	Likely evidence? (Yes/No)
12.		
13.		
14.		
15.		
16.		
17.		
18.		
19.		
20.		

Concluding Questions

1. Which pieces of garbage were most likely to link the suspect to a crime? Why?

____

____

____

____

Name ____________________ Date ____________________

14. Searching Through Garbage

STUDENT ACTIVITY PAGE

2. Were any pieces of evidence likely to contain DNA or fingerprints? If so, which ones?

3. What pieces of garbage were likely to be just regular garbage not connected to a crime? Why?

4. What evidence might have been on the bag itself?

Follow-Up Activity

There are very strict laws and guidelines about how and where garbage can be collected as evidence. Those rules vary from state to state. Research what laws your state has that relate to the collection of garbage as evidence. In what circumstances would an investigator need a search warrant?

15. Shoe Prints

INSTRUCTIONAL OBJECTIVES

Students will be able to:

- compare various shoe prints to identify similar structures
- analyze the structure of a shoe print
- determine if shoe prints match or not

NATIONAL SCIENCE EDUCATION STANDARDS CORRELATIONS

GRADES 5–8

Content standard	Bullet number	Content description	Bullet number(s)
A	1	Abilities necessary to do scientific inquiry	1–7
A	2	Understandings about scientific inquiry	1–5
E	1	Abilities of technological design	1–5
E	2	Understandings about science and technology	4–5
G	1	Science as a human endeavor	1–2

GRADES 9–12

Content standard	Bullet number	Content description	Bullet number(s)
A	1	Abilities necessary to do scientific inquiry	1–5
A	2	Understandings about scientific inquiry	3, 5
E	1	Abilities of technological design	1–4
G	1	Science as a human endeavor	2

VOCABULARY

- **footprint:** distinctive impression left behind by the human foot
- **shoe print:** distinctive impression left behind by the bottom of a shoe

MATERIALS

- inkless foot/shoe print kit
- various similar shoes (three basketball sneakers, three work boots, and so forth)
- magnifying glass
- ruler
- gloves (latex and nitrile)

15. Shoe Prints

TEACHER RESOURCE PAGE

HELPFUL HINTS AND DISCUSSION

Time frame: one class period
Structure: groups
Location: classroom

This lab requires the use of inkless foot/shoe print kits. These can be ordered from many of the popular science supply catalogs or from online dealers. Although the price of these kits tends to be high, most of them include enough materials for at least 100 tests. You may wish to review the package directions with students ahead of time, or rewrite the directions in student-friendly language on the board or overhead.

Students will be taking prints from three shoes and trying to match one of the prints they collected with a print that you will supply them with. This means that you must make prints beforehand and be sure that each group receives a premade print that will match one of the ones they are making.

Some students might be allergic to latex gloves, so be sure to have nitrile gloves available.

MEETING THE NEEDS OF DIVERSE LEARNERS

You might find that students with different abilities will benefit from extra help or extra challenges. Students who need extra challenges should complete the Extension Option and the Follow-Up Activity. The process for collecting an inkless shoe print must be followed very carefully. These students should be responsible for directing the collection. Students who need extra help might find the instructions for collecting a shoe print with the inkless printing system complicated. These students should carefully review the guidelines. Partnering with a high-achieving student might also be beneficial.

SCORING RUBRIC

Students meet the standard for this activity by:

- correctly using the foot/shoe print kit to collect data
- using a clear strategy for comparing prints
- forming conclusions that are based on the collected data
- appropriately using the magnifying glass to find details

RECOMMENDED INTERNET SITES

- **eNotes.com—World of Forensic Science: Shoeprints**
  www.enotes.com/forensic-science/shoeprints
- **Forensic-Evidence.com—Can Shoes Catch a Culprit? or Does a Shoeprint Lie?**
  http://forensic-evidence.com/site/ID/Shoeprint.html

ANSWER KEY

1. Answers will vary depending on the prints provided.
2. The measurements of length and width should be very close. Students should measure the prints themselves and not the shoes as this gives a more consistent comparison.
3. Answers will vary. If an investigator knows the manufacturer, he or she might be able to find out where the shoe was sold and link that purchase to the suspect.
4. A three-dimensional print would contain more information. It would show the depth of the grooves and might even suggest a height and weight for the suspect.

Name ______________________________ Date ____________________

15. Shoe Prints

STUDENT ACTIVITY PAGE

Objective

To understand how shoe prints can be used as evidence in a criminal investigation

Before You Begin

Shoes can leave behind prints just as your fingers can. The obvious prints, such as those left in paint, blood, or firm mud, are easily found by investigators. Other prints can be harder to collect. **Footprints** or **shoe prints** left in a carpet might seem too fragile to collect, but there are a number of ways they are preserved by modern crime-scene investigators. Careful use of high-intensity photography lights can bring out images. After pictures are taken, they can be magnified to life-size or larger for comparison with other collected prints or with shoes from suspects. Sometimes a device that creates a large static charge can draw particles up from the rug and onto a plastic sheet. There they can be photographed or lifted with a tape similar to fingerprint-lifting tape. Some fragile prints are sprayed with fixative sprays that help strengthen them. Then a cast is taken with plaster of Paris or dental-casting material.

Once a print is collected, investigators can use the print to identify the manufacturer of the shoe and the shoe size. This, in turn, allows police to look for a specific brand of shoe or to find where such a shoe might have been sold. They can also use the shoe size to estimate the size of the suspect. The depth to which a shoe sank in mud and the distance between shoe prints can help police determine the rough weight and height of a suspect. Further inspection of the print will reveal details such as nicks, cuts, and holes. Such details can be used to connect a specific print with a specific shoe.

Materials

- inkless foot/shoe print kit
- various similar shoes (three basketball sneakers, three work boots, and so forth)
- magnifying glass
- ruler
- gloves

Procedure

1. Put on your gloves to protect both you and the evidence.
2. Collect three shoes from your teacher.
3. Follow the instructions with the inkless foot/shoe print kit carefully. Each kit is a little different, and you must follow the instructions exactly or they will not work.
4. Be sure you use firm pressure when collecting a print so you can collect as much of the pattern on the bottom of the shoe as possible.
5. Be sure to label your prints as Shoe 1, Shoe 2, and Shoe 3 as you take them.

Name ______________________ Date ______________________

15. Shoe Prints

STUDENT ACTIVITY PAGE

6. Measure the width and length of each shoe print (not the actual shoe itself). Measure the widest width and longest length you can find.
7. Describe any distinguishing patterns you see in the print. Are there circles? Is there visible damage to the print? Is part of a heel, toe, or side worn smooth? Use the magnifying glass to look for small imperfections.
8. Look for manufacturer names or symbols and be sure to write them down.
9. Get the suspect shoe print you are trying to match from your teacher.
10. Repeat steps 6 and 7 for the new print.

EXTENSION OPTION

Trade matching prints with another group. Have them select some key points that indicate a match. Compare the points you chose to the points they chose and discuss the relative merit of each set of choices.

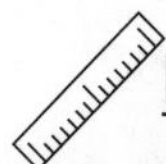

DATA COLLECTION AND ANALYSIS

	Length	Width	Distinguishing patterns	Manufacturer symbol or name
Shoe 1				
Shoe 2				
Shoe 3				
Suspect print				

Name ____________________ Date ____________________

15. Shoe Prints

STUDENT ACTIVITY PAGE

Concluding Questions

1. Which shoe was the match to the suspect print?

2. Although the print might have been an obvious match, you would be required to provide data if you needed to prove the match in court. How close were your measurements of length and width between the matching samples?

3. Did any of the prints have marks that would match them to a manufacturer? What could you do with this type of evidence?

4. If one of these shoe prints had been left in damp soil, a cast could have been made. What advantages might a cast have over a flat print?

Follow-Up Activity

Shoe prints are not only collected by inkless printing kits. Research some other methods for collecting shoe prints. Make a list of the many surfaces on which a suspect might leave a shoe print. What would be the best method for collecting a print from each of the various surfaces?

16. Tool Marks

TEACHER RESOURCE PAGE

Instructional Objectives

Students will be able to:

- analyze the structure of a tool mark
- determine if tool marks match or not
- compare various tool marks to identify similar structures

National Science Education Standards Correlations

Grades 5–8

Content standard	Bullet number	Content description	Bullet number(s)
A	1	Abilities necessary to do scientific inquiry	1–7
A	2	Understandings about scientific inquiry	1–5
E	1	Abilities of technological design	1–5
E	2	Understandings about science and technology	4–5
G	1	Science as a human endeavor	1–2

Grades 9–12

Content standard	Bullet number	Content description	Bullet number(s)
A	1	Abilities necessary to do scientific inquiry	1–5
A	2	Understandings about scientific inquiry	3, 5
E	1	Abilities of technological design	1–4
G	1	Science as a human endeavor	2

Vocabulary

- **tool mark:** any of the various impressions left behind by the use of a tool such as a hammer, screwdriver, knife, wrench, and so forth

Materials

- copper pipe
- block of wood (one per group)
- pliers
- screwdrivers
- hammer
- goggles
- magnifying glass

16. Tool Marks

Helpful Hints and Discussion

Time frame: one class period
Structure: partners or groups
Location: classroom

You will need to use pliers to pre-mark some lengths of copper pipe and lightly hammer a screwdriver into each wooden block to leave the tool marks. Be sure that when students collect their pre-marked items, they also receive the matching pliers and screwdrivers that made the marks. It would be helpful if students received pliers that were similar in size but with slightly different teeth. The screwdrivers should not be very different in size, but they should be different enough so that it is clear which one the marks came from.

Meeting the Needs of Diverse Learners

You might find that students with different abilities will benefit from extra help or extra challenges. Students who need extra challenges should complete the Extension Option and the Follow-Up Activity. These students should be encouraged to use the magnifying glass very carefully to include or exclude all of the samples and not just the suspected match. Students who need extra help should be made aware of the safety considerations when using the hammer and when handling the other tools. They will also need extra instruction on the importance of not damaging the original evidence with the suspect tools.

Scoring Rubric

Students meet the standard for this activity by:

- following the procedure carefully
- using the magnifying glass to increase the quality of collected evidence
- carefully collecting and recording data
- forming a conclusion based on the collected data

Recommended Internet Sites

- **Missouri State Highway Patrol—Toolmarks**
  www.mshp.dps.mo.gov/MSHPWeb/PatrolDivisions/CLD/Firearms/toolmarks.html
- **Suite101.com—Toolmarks at a Crime Scene**
  http://crime-scene-processing.suite101.com/article.cfm/toolmarks_at_a_crime_scene

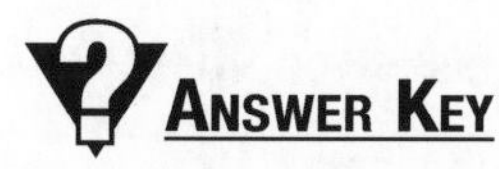

Answer Key

1. Answers will vary, but the spacing of the teeth and width of the sample are common responses.
2. Answers will vary, but students can usually tell by the width of the mark and the fit of the suspect tool in the hole.
3. If the original sample is touched with the suspect tools, it could be damaged and then excluded as evidence. It would be safer to make a cast of the tool marks or to match them photographically.

Name ______________________________ Date ______________________________

16. Tool Marks

STUDENT ACTIVITY PAGE

OBJECTIVE

To use tool marks as evidence to connect a tool owned by a suspect to a crime

BEFORE YOU BEGIN

You're a criminal. You're outside a house. You want to be inside the house. You walk around the building checking for open doors or open windows, but eventually you realize there's no easy way in. You see a laptop and a digital camera sitting on a table just inside, and you have to have them. You reach into your backpack, take out a large screwdriver, and try to pry the door. Nothing. You have a small pry bar, and with a little work, you eventually pry part of the door far enough so that it unlatches. You're in, and the goods are going into your backpack.

Later, you're sitting at home when there's a knock at the door. It's the police. You didn't leave any DNA or fingerprints, and you're pretty sure you didn't leave any footprints on the concrete walkway around the house. However, someone saw you, and the police recognized you from the description. Luckily for you, you left the laptop and camera at a friend's house. So you're in the clear, right?

The police have a search warrant, and they find your backpack. They find the screwdriver and the pry bar. They look at some pictures and at your tools. Then they put the handcuffs on you. The kind of **tool marks** found around the door frame tell police they are looking for a screwdriver—they know the exact size of the end. They are also looking for a pry bar that has a big knick on the end, just like the knick you put in your pry bar when you tried to drive nails with it. Oops.

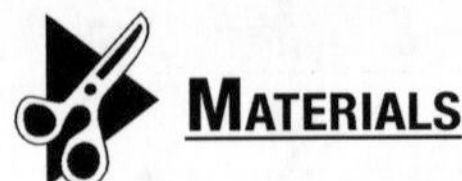

MATERIALS

- copper pipe
- block of wood
- pliers
- screwdrivers
- hammer
- goggles
- magnifying glass

PROCEDURE

1. Put on your safety goggles.
2. Collect a piece of copper pipe from your teacher that has tool marks from a pair of pliers. Also collect the three pairs of pliers that were gathered as evidence.
3. Find the tool marks on the pipe and look at them carefully with the magnifying glass. Record any distinguishing features you notice.
4. Look at your sample pliers. Try to line up the teeth marks of the pliers with the marks in the pipe. Be careful not to damage the original marks. Try all three pairs until you think you have a match.
5. When you decide on a match, use the magnifying glass to look at both the tool marks and the actual pliers. Try to find any specific marks that might be unique to the tool (such as damage to the teeth).

Name ______________________________ Date ______________________________

16. Tool Marks

STUDENT ACTIVITY PAGE

6. Collect the wooden block with the tool mark made by a screwdriver. Collect the screwdrivers that were gathered as evidence.
7. Look at the mark in the wood with your magnifying glass and record any distinguishing features.
8. Carefully try to fit the evidence screwdrivers into the hole in the wooden block one at a time. Be careful not to damage the original tool mark.
9. When you think you have a match, look at the screwdriver and the hole again with the magnifying glass to see if there are any specific marks that might be unique to the tool.
10. When you have recorded your data, use the hammer to carefully tap the screwdriver you think is a match into the wood in a place at least 5 centimeters from the original mark. Determine if the two holes look the same.

Extension Option

Make tool marks with all of the sample tools you have and see if any of them could also be a match other than the one you decided on.

Data Collection and Analysis

1. Distinguishing characteristics on suspect pliers:

____

____

____

____

2. Common characteristics between tool mark and suspect pliers:

____

____

____

____

3. Distinguishing characteristics on suspect screwdriver:

____

____

____

____

Name ______________________________ Date ______________________________

16. Tool Marks

STUDENT ACTIVITY PAGE

4. Common characteristics between tool mark and suspect screwdriver:

____

____

____

____

CONCLUDING QUESTIONS

1. Was one of the pliers an obvious match to the tool mark on the copper pipe? If so, what characteristics of the tool mark made the match obvious?

____

____

____

____

2. Was one of the screwdrivers an obvious match to the tool mark on the wooden block? If so, what characteristic of the tool mark made the match obvious?

____

____

____

____

3. What problems might arise from allowing suspect tools to touch the original evidence? How might an investigator avoid such problems?

____

____

____

____

FOLLOW-UP ACTIVITY

There are a number of databases that keep track of the various kinds of tools and the marks they leave. Research the kinds of tool marks that have been used by forensic scientists and make a complete list of them.

17. Microscopic Fibers

TEACHER RESOURCE PAGE

INSTRUCTIONAL OBJECTIVES

Students will be able to:

- identify natural and synthetic fibers
- compare fiber samples
- describe characteristics of fibers
- use a microscope to analyze fibers

NATIONAL SCIENCE EDUCATION STANDARDS CORRELATIONS

GRADES 5–8

Content standard	Bullet number	Content description	Bullet number(s)
A	1	Abilities necessary to do scientific inquiry	1–7
A	2	Understandings about scientific inquiry	1–5
E	1	Abilities of technological design	1–5
E	2	Understandings about science and technology	4–5
G	1	Science as a human endeavor	1–2

GRADES 9–12

Content standard	Bullet number	Content description	Bullet number(s)
A	1	Abilities necessary to do scientific inquiry	1–5
A	2	Understandings about scientific inquiry	3, 5
B	2	Structure and properties of matter	4
E	1	Abilities of technological design	1–4
G	1	Science as a human endeavor	2

VOCABULARY

- **fiber:** generally a small sample of material that can be combined with other fibers to form threads, yarns, and other fabrics
- **naturally occurring:** describes a material directly from an animal or a plant source
- **synthetic:** created by artificial means

17. Microscopic Fibers

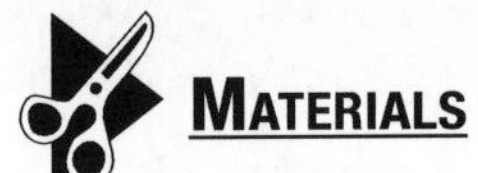

MATERIALS

- fiber samples cut into small squares:
  - cotton
  - wool
  - polyester
  - rayon
  - silk
  - blended sample (rayon and cotton, for example)
- tweezers
- scissors
- microscope
- color chart (found at most paint-supply stores)

HELPFUL HINTS AND DISCUSSION

Time frame: one class period
Structure: individuals or partners
Location: classroom

You will need to provide students with labeled examples of all the fibers listed in the materials list. A good way to keep the samples organized is to put them in small resealable plastic bags that each have a label on the outside. The two unknowns can be any of the six samples, and it is a good idea if each individual or group has a different combination of unknowns. Some fibers do not cut well and might need to be cut with a razor blade. For safety reasons, you might want to do such cutting instead of letting students use razor blades.

MEETING THE NEEDS OF DIVERSE LEARNERS

You might find that students with different abilities will benefit from extra help or extra challenges. Students who need extra challenges should complete the Extension Option and the Follow-Up Activity. These students should be encouraged to use technical terms to describe the samples when they are viewed under magnification. Students who need extra help might have a difficult time discerning the pattern of the blended sample. These students might benefit from working with a high-achieving student.

SCORING RUBRIC

Students meet the standard for this activity by:

- correctly using the microscope to view fibers
- correctly recording color data and sketches
- making comparisons based on observations
- understanding basic differences among fiber samples

17. Microscopic Fibers

TEACHER RESOURCE PAGE

RECOMMENDED INTERNET SITES

- **Federal Bureau of Investigation—Forensic Science Communications: Hairs, Fibers, Crime, and Evidence**
  www.fbi.gov/hq/lab/fsc/backissu/july2000/deedric3.htm
- **Home Furnishing Trade Marketplace—What is a Fiber?**
  www.textilefurnishings.com/what-is-fiber.html

ANSWER KEY

1. Answers will vary based on sample selection.
2. Generally color and cross section are strong suggestions for a match. Students might not notice the overall structure as well when looking at a sample.
3. Answers will vary based on sample selection.
4. Generally color and cross section are strong suggestions for a match. Students might not notice the overall structure as well when looking at a sample.
5. With a large fiber sample, investigators are more likely to find out where and how it was manufactured. This might allow them to track where the original item was sold or who it was sold to. This also gives them a large piece of evidence that might be recognized as belonging to a damaged item owned by a suspect.
6. Answers will vary based on sample selection.

Name ______________________________ Date ______________________

17. Microscopic Fibers

STUDENT ACTIVITY PAGE

OBJECTIVE

To use a microscope to view and identify various types of fibers

BEFORE YOU BEGIN

Fibers can be very small. They are easily dislodged from objects during violent actions. If a man wearing a wool sweater is attacked by another man, it is possible that the attacker will end up with wool fibers under his fingernails or on his clothing. Sometimes such evidence is too small to be picked up with fingers or tweezers. If so, it can be collected with tape or with a special vacuum that has a filter for catching small fibers.

There are many different kinds of fibers. Fibers can be split into two large groups: **synthetic** and **naturally occurring.** The naturally occurring fibers can come from a mineral source such as asbestos. They can also come from a plant or an animal source. Wood and cotton are both examples of naturally occurring plant fibers. Animal hair and wool are both examples naturally occurring animal fibers.

Synthetic fibers come from many sources. Clothing, upholstery, linens, carpeting, ropes, and even toys are all made from synthetic materials. These materials include rayon, nylon, Dacron, acrylic, polyester, spandex, and orlon.

All of these fibers can be identified by a number of properties. Sometimes all that is needed is to look at the fibers under a microscope. Sometimes samples have to be tested and identified by chemical means. Some types of advanced testing use devices such as the liquid-gas chromatograph. Heating, burning, or shining light through a sample of fiber produces a unique result. This result can be compared to a database that contains information on thousands of kinds of fibers, both synthetic and natural.

MATERIALS

- fiber samples cut into small squares:
  - cotton
  - wool
  - polyester
  - rayon
  - silk
  - blended sample
- tweezers
- scissors
- microscope
- color chart

Name ____________________ Date ____________________

17. Microscopic Fibers

STUDENT ACTIVITY PAGE

PROCEDURE

1. Collect the known samples from your teacher. Be sure to keep them separate and to only take one from its container at a time.
2. Look at each sample under the highest magnification provided by your microscope. Draw a sketch of what you see. Try to match the color with the color chart.
3. If possible, get a single strand of each fiber with the tweezers and cut it with the scissors. If you can see the end, draw a sketch of the cross section.
4. Draw a sketch of the large piece of the blended sample. If possible, draw the pattern of the weave.
5. Collect the two unknown samples from your teacher.
6. Try to match the unknowns by looking at them under the microscope and identifying structure and color.
7. Cut an individual strand of each unknown and draw the cross section.

EXTENSION OPTION

Look at some samples of the clothing that you or a classmate is wearing. See if you can identify different weave patterns in the fabric.

DATA COLLECTION AND ANALYSIS

	Sketch	Color	Cross-section sketch
Cotton			
Wool			
Polyester			

(chart continues on next page)

Name ______________________________ Date ______________________

17. Microscopic Fibers

STUDENT ACTIVITY PAGE

	Sketch	Color	Cross-section sketch
Rayon			
Silk			
Blended sample			
Unknown 1			
Unknown 2			

Concluding Questions

1. Which fiber was a match for unknown 1?

____

____

2. What characteristics of unknown 1 helped you make the match?

____

____

____

____

Name ____________________ Date ____________________

17. Microscopic Fibers

STUDENT ACTIVITY PAGE

3. Which fiber was a match for unknown 2?

4. What characteristics of unknown 2 helped you make the match?

5. What advantages do investigators have if they find a large piece of fiber evidence such as your blended sample?

6. Which fibers were natural and which fibers were synthetic?

FOLLOW-UP ACTIVITY

There are a large number of weaving patterns. Such patterns can help investigators identify where and how a cloth sample was made. Research and find five kinds of weave, and draw a sketch of each kind.

18. Human Hair versus Animal Hair

TEACHER RESOURCE PAGE

Instructional Objectives

Students will be able to:

- distinguish human hair from animal hair
- describe human hair and animal hair
- identify a sample of hair as human or non-human
- use a microscope to analyze human hair and animal hair

National Science Education Standards Correlations

Grades 5–8

Content standard	Bullet number	Content description	Bullet number(s)
A	1	Abilities necessary to do scientific inquiry	1–7
A	2	Understandings about scientific inquiry	1–5
C	2	Reproduction and heredity	4–5
C	5	Diversity and adaptations of organisms	1
E	1	Abilities of technological design	1–5
E	2	Understandings about science and technology	4–5
G	1	Science as a human endeavor	1–2

Grades 9–12

Content standard	Bullet number	Content description	Bullet number(s)
A	1	Abilities necessary to do scientific inquiry	1–5
A	2	Understandings about scientific inquiry	3, 5
C	2	Molecular basis of heredity	1–2
E	1	Abilities of technological design	1–4
G	1	Science as a human endeavor	2

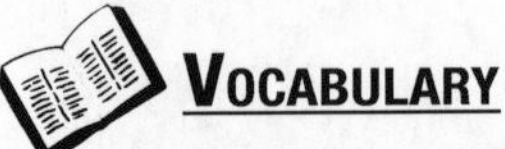

Vocabulary

- **hair:** threadlike growth found on the surface of humans and many animals

18. Human Hair versus Animal Hair

TEACHER RESOURCE PAGE

MATERIALS

- gloves (latex and nitrile)
- human hair samples
- animal hair samples
- microscope
- tweezers
- color chart (found at most paint-supply stores)
- scissors

HELPFUL HINTS AND DISCUSSION

Time frame: one class period
Structure: individuals or partners
Location: classroom

Students will need a wide variety of hair samples to view under the microscope. Students may choose to supply hair samples for other students to view. The animal samples should be collected from animals that are as different as possible. Cats and dogs are common sources, but you might want to ask students to provide hair from more unusual animals such as ferrets, guinea pigs, horses, or any other animals they have access to. The unknown hairs should match the known samples that the students have already received.

Some students might be allergic to latex gloves, so be sure to have nitrile gloves available.

MEETING THE NEEDS OF DIVERSE LEARNERS

You might find that students with different abilities will benefit from extra help or extra challenges. Students who need extra challenges should complete the Extension Option and the Follow-Up Activity. These students should also be encouraged to examine other hair samples from their own bodies. Students who need extra help might have difficulty sketching the structures they see under the microscope. You might wish to take extra time to assist them in transferring the image they see to the paper.

SCORING RUBRIC

Students meet the standard for this activity by:

- correctly using the microscope to view hairs
- correctly recording color data and sketches
- making comparisons based on observations
- understanding basic differences between animal hair and human hair

18. Human Hair versus Animal Hair

TEACHER RESOURCE PAGE

RECOMMENDED INTERNET SITES

- **Federal Bureau of Investigation—Forensic Science Communications: Forensic Human Hair Examination Guidelines**
  www.fbi.gov/hq/lab/fsc/backissu/april2005/standards/2005_04_standards02.htm
- **Federal Bureau of Investigation—Forensic Science Communications: Hairs, Fibers, Crime, and Evidence**
  www.fbi.gov/hq/lab/fsc/backissu/july2000/deedric1.htm

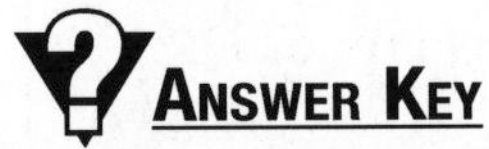

ANSWER KEY

Answers will vary based on the hair samples students receive. However, students should be able to identify some of the major differences between human hair and animal hair.

Name ______________________ Date ______________________

18. Human Hair versus Animal Hair

STUDENT ACTIVITY PAGE

OBJECTIVE

To identify some of the differences between human hair and animal hair

BEFORE YOU BEGIN

Arriving at a crime scene, investigators are faced with the task of finding where to begin. Once they have chosen a location, they know that there are many kinds of evidence to collect. Some things are large and obvious, such as guns or bloody clothing. Other things that might seem insignificant could be far more useful later.

A dead body is found in a house. The person has been strangled. Did the killer use a rope? Did the killer use his or her bare hands? Was the murder weapon a piece of wire or rope? While looking at the body, an investigator notices that the victim's sweatshirt is covered with fibers. Are they natural, or are they synthetic? Closer inspection leads the investigator to conclude that they are very fine **hairs,** both long and short. Did they come from a man or a woman? Is that one an eyelash? Is it an eyebrow hair? Could they be from a beard? With more searching, more kinds of hairs are found. Is that one from a wig? Are those short, fine hairs from a cat or from a ferret? Could the long ones possibly be from a horse? Although it seems unlikely, a horse hair would be a very unusual piece of evidence. It would certainly narrow down the suspects faster than if it was from a more common animal such as a cat or dog.

When animal hairs and human hairs are viewed under a microscope, there are obvious differences between them. At a level even smaller than the simple microscope can see, DNA can tell investigators about the kind of hair they have found. Investigators can even use DNA to narrow down the unique animal or human the hair belongs to.

MATERIALS

- human hair samples
- animal hair samples
- microscope
- tweezers
- color chart
- scissors
- gloves

PROCEDURE

1. Put on your gloves to protect both you and the evidence.
2. Collect the samples of human hair provided by your teacher. Only remove one sample at a time from its container.
3. View the first human hair sample under a microscope. Draw a sketch of what you see.
4. Find the best match for the color of the hair on the color chart and record it in the data table.

Name ______________________________ Date ______________________

18. Human Hair versus Animal Hair

STUDENT ACTIVITY PAGE

5. Carefully cut a piece of the first sample of human hair with a pair of scissors and view the cross section under the microscope. Draw a sketch of what you see.
6. Repeat steps 3–5 for the other two human hair samples. Then return your samples to your teacher.
7. Collect the animal hairs provided by your teacher. Only remove one sample at a time from its container.
8. View the first animal hair under a microscope. Draw a sketch of what you see.
9. Find the best match for the color of the hair on the color chart and record it in the data table.
10. Carefully cut a piece of the first sample of animal hair with a pair of scissors and view the cross section under the microscope. Draw a sketch of what you see.
11. Repeat steps 8–10 for the other two animal hair samples. Then return your samples to your teacher.
12. Collect the unknown hair from your teacher.
13. View the first unknown hair under a microscope. Draw a sketch of what you see.
14. Find the best match for the color of the hair on the color chart and record it in the data table.
15. Carefully cut the unknown hair with a pair of scissors and view the cross section under the microscope. Draw a sketch of what you see.
16. Repeat steps 13–15 for the second unknown hair.

EXTENSION OPTION

View your own hair under the microscope. Compare the structures of hair from your head, eyebrows, arms, legs, or eyelashes.

DATA COLLECTION AND ANALYSIS

	Overall sketch	Color match	Cross-section sketch
Human hair 1			
Human hair 2			

(chart continues on next page)

Name ______________________ Date ______________________

18. Human Hair versus Animal Hair

STUDENT ACTIVITY PAGE

	Overall sketch	Color match	Cross-section sketch
Human hair 3			
Animal hair 1			
Animal hair 2			
Animal hair 3			
Unknown hair 1			
Unknown hair 2			

Name ______________________________ Date ______________________________

18. Human Hair versus Animal Hair

Student Activity Page

Concluding Questions

1. Did the human hairs share any similar characteristics? If so, what were they?

____

____

____

____

2. Did the animal hairs share any similar characteristics? If so, what were they?

____

____

____

____

3. What kind of hair was unknown hair 1? What characteristics led you to that conclusion?

____

____

____

____

4. What kind of hair was unknown hair 2? What characteristics led you to that conclusion?

____

____

____

____

Follow-Up Activity

The FBI has an extremely long list of descriptions that are used to identify the various structures of hair. Research to find out what all of these descriptions are and what they mean.

19. Summary: A Car As a Crime Scene

INSTRUCTIONAL OBJECTIVES

Students will be able to:

- evaluate a large collection of evidence
- conduct an investigation with many parts
- identify different areas of investigation
- connect evidence

NATIONAL SCIENCE EDUCATION STANDARDS CORRELATIONS

GRADES 5–8

Content standard	Bullet number	Content description	Bullet number(s)
A	1	Abilities necessary to do scientific inquiry	1–7
A	2	Understandings about scientific inquiry	1–5
E	1	Abilities of technological design	1–5
E	2	Understandings about science and technology	4–5
G	1	Science as a human endeavor	1–2

GRADES 9–12

Content standard	Bullet number	Content description	Bullet number(s)
A	1	Abilities necessary to do scientific inquiry	1–5
A	2	Understandings about scientific inquiry	3, 5
B	4	Motions and forces	1
E	1	Abilities of technological design	1–4
E	2	Understandings about science and technology	3
G	1	Science as a human endeavor	2

VOCABULARY

- **Federal Bureau of Investigation (FBI):** an organization that deals with crimes at the federal level
- **protocol:** the correct order for processing a crime scene

MATERIALS

- car (cleaned and vacuumed beforehand)
- digital camera
- gloves (latex and nitrile)
- fingerprint dusting kit
- plastic bags
- tweezers
- microscope
- meterstick
- tape measure
- scissors
- magnifying glass
- tray
- scale
- graduated cylinder
- string or crime-scene tape

HELPFUL HINTS AND DISCUSSION

Time frame: two class periods
Structure: groups
Location: vehicle

This lab will incorporate techniques that students used from the following earlier activities:

- Photographing a Crime Scene
- Processing a Crime Scene
- Fingerprints
- Car Accident
- Density of Glass Fragments
- Searching Through Garbage
- Human Hair versus Animal Hair

The complexity of the crime scene is only dependent on the time and resources you have in creating it. The general idea is that an abandoned car has been found (Crime Scene 1), and investigators suspect either foul play or that the car has been used in the commission of a crime. You will have information collected from another crime scene (Crime Scene 2, an imaginary crime scene that you will provide "evidence" from). This information may be connected by the various data and tests the students perform on evidence from the vehicle.

1. Students will need to photograph the crime scene (Crime Scene 1) and decide how to search the vehicle. You will need to plant fingerprints for the students to collect. These fingerprints will be compared to the ones you supplied from Crime Scene 2.
2. You will need to indicate the length of skid marks left by the car as it came to a stop. You should provide a description of the car and an estimate, supplied by a witness, of how fast the car was going.

(continued on next page)

3. You will need to plant glass fragments that students will have to find the density of so they can match the glass in the car with glass found at another crime scene. These fragments could be from a broken window at Crime Scene 2. You can provide the density of the glass from the window.
4. You can plant garbage in the car such as crumpled up ransom notes, receipts, or handwriting samples that can be compared to information from Crime Scene 2.
5. You will also need to plant some hairs in a conspicuous place so students can test them to see if they are from an animal or a human. The suspect from Crime Scene 2 had a dog with him, and students will need to test the hair to see if it is human or not.

Reading through the concluding questions will help you decide what evidence to plant and how to plant it. Some students might be allergic to latex gloves, so be sure to have nitrile gloves available.

Meeting the Needs of Diverse Learners

You might find that students with different abilities will benefit from extra help or extra challenges. Students who need extra challenges should complete the Extension Option and the Follow-Up Activity. These students should be group leaders. There is a lot of information to gather and much evidence that could be missed. These students should be allowed to organize all of the collected evidence. Students who need extra help should be reminded of the earlier activities and how each one was carried out. Copies of the procedures from earlier activities will be beneficial to these students.

Scoring Rubric

Students meet the standard for this activity by:

- following procedures for processing a crime scene
- correctly using tools and equipment
- properly recording data
- drawing conclusions that are supported by the evidence
- understanding the scientific value of each kind of tool, each piece of evidence, and each procedure used to test, collect, or record data

Recommended Internet Sites

- **Crime-Scene-Investigator.net—Crime Scene Response**
  www.crime-scene-investigator.net/csi-response.html
- **Crime-Scene-Investigator.net—Evidence Collection**
  www.crime-scene-investigator.net/csi-collection.html
- **HowStuffWorks—How Crime Scene Investigation Works**
  http://science.howstuffworks.com/csi.htm

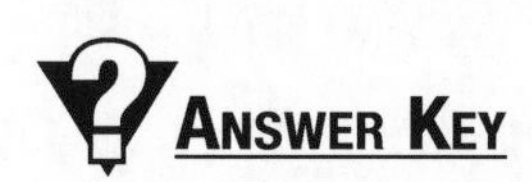

Answer Key

The answers supplied by students will depend entirely on the nature of the evidence supplied by the teacher.

Name ______________________________ Date ______________

19. Summary: A Car As a Crime Scene

STUDENT ACTIVITY PAGE

OBJECTIVE

To use a variety of skills to evaluate a crime scene with multiple sources of evidence

BEFORE YOU BEGIN

Crime-scene investigators will tell you that they rely heavily on **protocol** when they process a crime scene. The amount of evidence that might be scattered around a ten-car pileup on a major highway, or in a large home where three victims were murdered in three different rooms, can be overwhelming. Some investigators use suggestions from the **FBI,** the **Federal Bureau of Investigation.** Others must follow the procedure they receive from state or local officials. In any case, processing a scene requires patience and careful attention to detail.

Imagine that a car has skidded to a stop on the side of the road. The keys are still in it, and the motor ran until there was no more gas. No one knows where the car came from. However, a passing motorist noticed that both the front doors were open but nobody was in the vehicle. A quick check by a police officer reveals that the car belongs to a person who has been missing for three days. You are the crime-scene investigator sent to process the scene—what do you do?

Think about the kinds of evidence have you read about that could be used to determine what happened in the car. You need to secure the area, take photographs, and look for fingerprints, blood, and blood patterns. Was this car stopped because of an accident? Is there damage to the car, broken glass, or tire tracks that might suggest what happened? What's in the car? Is there soil on the floor, a ransom note, or garbage that might help your search? Are there footprints on the floor mats or tool marks left from a weapon? Can you see any fibers that could be artificial or synthetic? Your ability to collect and interpret the evidence can help solve the mystery.

MATERIALS

- car
- digital camera
- gloves
- fingerprint dusting kit
- plastic bags
- tweezers
- microscope
- meterstick
- tape measure
- scissors
- magnifying glass
- tray
- scale
- graduated cylinder
- string or crime-scene tape

Name ______________________________ Date ______________________

19. Summary: A Car As a Crime Scene

STUDENT ACTIVITY PAGE

Procedure

1. Put on your gloves to protect you and the evidence.
2. Determine the boundaries of the crime scene. Mark the scene off with tape or string to indicate the boundaries.
3. Take photographs of the crime scene and surroundings. Be sure to get pictures of the overall setting, the middle range, and close-ups.
4. Determine the best method to search the crime scene. An area of the crime scene will be assigned to each group. Make a sketch of the scene, and indicate on the sketch where each piece of evidence is found.
5. Collect any evidence that needs to be tested. Record what it is, its basic characteristics, and where it was found.
6. Test your various samples, and compare them to the samples provided by your teacher.

Extension Option

As a group, present the evidence you found at the crime scene. Be sure to compare evidence collected at Crime Scene 1 with evidence from Crime Scene 2.

Data Collection and Analysis

Crime Scene Sketch

Name ______________________ Date ______________

19. Summary: A Car As a Crime Scene

STUDENT ACTIVITY PAGE

Evidence found	Description	Location collected
1.		
2.		
3.		
4.		
5.		
6.		
7.		
8.		
9.		
10.		

(chart continues on next page)

Name ______________________ Date ______________________

19. Summary: A Car As a Crime Scene

STUDENT ACTIVITY PAGE

Evidence found	Description	Location collected
11.		
12.		
13.		
14.		
15.		
16.		
17.		
18.		
19.		
20.		

Name ______________________ Date ______________________

19. Summary: A Car As a Crime Scene

STUDENT ACTIVITY PAGE

CONCLUDING QUESTIONS

1. Did you find any fingerprints at Crime Scene 1 that matched prints from Crime Scene 2?

2. How fast was the car going when it skidded to a halt? Does this speed closely match the speed supplied by the witness?

3. Did the density of the glass fragments match the density of the glass from the window that was broken at Crime Scene 2?

4. Did any of the collected garbage connect the car to the crime committed at Crime Scene 2? How?

Name ______________________ Date ______________________

19. Summary: A Car As a Crime Scene

STUDENT ACTIVITY PAGE

5. Was the hair collected animal or human? How could you tell?

____

____

____

____

6. Based on the evidence you collected, is it likely that the driver of this car was involved with the crime that was committed at Crime Scene 2?

____

____

____

____

FOLLOW-UP ACTIVITY

As a class, review the manner in which all of the evidence was collected and tested. What steps would have insured better organization for the collection and treatment of evidence?

20. Summary: Missing Person—Your Teacher!

TEACHER RESOURCE PAGE

INSTRUCTIONAL OBJECTIVES

Students will be able to:

- evaluate a large collection of evidence
- conduct an investigation with many parts
- identify different areas of investigation
- connect evidence

NATIONAL SCIENCE EDUCATION STANDARDS CORRELATIONS

GRADES 5–8

Content standard	Bullet number	Content description	Bullet number(s)
A	1	Abilities necessary to do scientific inquiry	1–7
A	2	Understandings about scientific inquiry	1–5
E	1	Abilities of technological design	1–5
E	2	Understandings about science and technology	4–5
G	1	Science as a human endeavor	1–2

GRADES 9–12

Content standard	Bullet number	Content description	Bullet number(s)
A	1	Abilities necessary to do scientific inquiry	1–5
A	2	Understandings about scientific inquiry	3, 5
E	1	Abilities of technological design	1–4
E	2	Understandings about science and technology	3
G	1	Science as a human endeavor	2

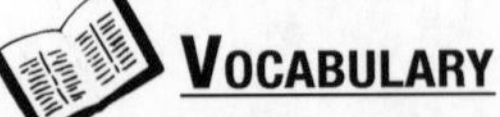

VOCABULARY

- **evidence:** materials collected at a crime scene that may be used in a court of law

20. Summary: Missing Person—Your Teacher!

TEACHER RESOURCE PAGE

MATERIALS

- gloves (latex and nitrile)
- goggles
- digital camera
- computer
- artificial blood
- sterile swabs
- distilled water
- rinse bottles
- small beakers, 50- or 100-ml
- luminol
- potassium hydroxide
- hydrogen peroxide (3% over-the-counter concentration)
- magnifying glass
- microscope
- tweezers
- scissors
- hair samples
- teacher handwriting sample
- pliers
- lockbox
- wire
- fiber samples:
  - cotton
  - wool
  - polyester
  - rayon
  - silk

HELPFUL HINTS AND DISCUSSION

Time frame: two class periods
Structure: groups
Location: classroom

This lab will incorporate techniques that students used from the following earlier activities:

- Photographing a Crime Scene
- Processing a Crime Scene
- Detecting Blood
- Blood Pattern Analysis
- Handwriting Analysis
- Tool Marks
- Microscopic Fibers

(continued on next page)

20. Summary: Missing Person—Your Teacher!

TEACHER RESOURCE PAGE

The concept of the activity is that students show up for class and discover that you, their teacher, are missing. You will, of course, be present, but students will have to interpret the evidence they find to determine if you have been the victim of foul play or an accident. How you present the evidence will indicate whether you were, for example, kidnapped, or if you cut your finger and went to see the school nurse. It is your choice. Some evidence will be presented as having been collected before students' arrival. Students should be supplied with evidence such as a sample of your hair, a sample of your handwriting, an assortment of pliers that might have been used to open the lockbox, and information regarding the kind of fiber in the shirt you were wearing when you were last seen.

1. You will have to leave fake blood at the scene. The blood will be tested using the luminol solution from Activity 4, Detecting Blood.
2. The blood should be left in a pattern that students can interpret. It can be pooled, spattered, or dropped, depending on the crime, or lack thereof, you wish to create. More dramatic spattering will suggest a more violent scenario.
3. A note saying that you stepped out should be left and signed with your signature. The note can be real or fake. A comparison sample of your handwriting should be made available.
4. You can provide a lockbox that is left open on the desk with a few coins in it, indicating that it held some money. A piece of wire that is wrapped around the box can be indented with tool marks from some pliers to indicate that it was torn open. Sample pliers can be supplied for students to attempt to make a match. Each pair should be marked as belonging to a particular staff member, perhaps with marker on the handle.
5. You can place some fibers around the lockbox that students can collect and identify. These can be saved to match to a suspect. Students can also identify what kind of fibers they are as they did in Activity 17, Microscopic Fibers. The kind of fibers in your shirt can be supplied to students. Fibers from suspects can also be supplied.

Reading through the concluding questions will help you decide what evidence to plant and how to plant it. Some students might be allergic to latex gloves, so be sure to have nitrile gloves available.

MEETING THE NEEDS OF DIVERSE LEARNERS

You might find that students with different abilities will benefit from extra help or extra challenges. Students who need extra challenges should complete the Extension Option and the Follow-Up Activity. These students will also make good group leaders. If Activity 19, Summary: A Car As a Crime Scene, was done earlier, students should use the organizational skills learned in that lab to gather evidence in this lab. Students who need extra help might need a review of the procedures for testing evidence. They will need to be reminded of the safety issues involved with handling luminol solution in particular.

20. Summary: Missing Person—Your Teacher!

SCORING RUBRIC

Students meet the standard for this activity by:

- following procedures for processing a crime scene
- correctly using tools and equipment
- properly recording data
- drawing conclusions that are supported by the evidence
- understanding the scientific value of each kind of tool, each piece of evidence, and each procedure used to test, collect, or record data

RECOMMENDED INTERNET SITES

- **Crime-Scene-Investigator.net—Crime Scene Response**
  www.crime-scene-investigator.net/csi-response.html
- **Crime-Scene-Investigator.net—Evidence Collection**
  www.crime-scene-investigator.net/csi-collection.html
- **HowStuffWorks—How Crime Scene Investigation Works**
  http://science.howstuffworks.com/csi.htm

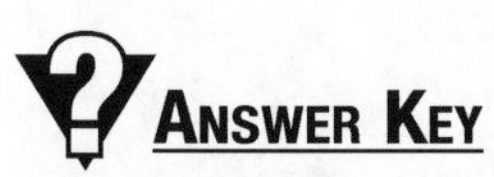

ANSWER KEY

Answers will vary depending on the scenario created by the teacher. All answers, however, should be supported by the forensics techniques students have learned.

Name ______________________ Date ______________________

20. Summary: Missing Person—Your Teacher!

STUDENT ACTIVITY PAGE

OBJECTIVE

To use a variety of skills to evaluate a crime scene with multiple sources of evidence

BEFORE YOU BEGIN

You walk into your classroom, but your teacher is not waiting for you. There are signs of a struggle and something that looks like blood on the desk. What has happened? Did your teacher get cut on a pair of scissors and go to find a bandage, or is something more sinister going on? Are there items missing that you would usually find in the classroom? Is your teacher mixed up in an international spy ring as a second job? What clues around the room might help you find your teacher? Think of all of the kinds of **evidence** that could be collected at the scene of a kidnapping. For legal purposes, evidence is information or items presented to prove or disprove an argument in court, and such evidence can take on many forms. Is there a ransom note? What things are of no importance, and what things could be valuable in your search? The classroom has desks, books, chairs, and maybe computers. What else is here? Which flat surfaces are likely to be covered with fingerprints or DNA evidence? What evidence could be on the floor that you have already walked through and ruined?

You are a crime-scene investigator, and it is your job to draw from all of the scientific skills you have learned to find out what happened to your teacher.

MATERIALS

- gloves
- goggles
- digital camera
- computer
- artificial blood
- sterile swabs
- distilled water
- rinse bottles
- small beakers, 50- or 100-ml
- luminol
- potassium hydroxide
- hydrogen peroxide
- magnifying glass
- microscope
- tweezers
- scissors
- hair samples
- teacher handwriting sample
- pliers
- lockbox
- wire
- fiber samples:
  - cotton
  - wool
  - polyester
  - rayon
  - silk

Name ________________________________ Date ________________________________

20. Summary: Missing Person—Your Teacher!

Student Activity Page

Procedure

1. Put on your gloves to protect you and the evidence.
2. Determine the boundaries of the crime scene. Mark the scene off with tape or string to indicate the boundaries.
3. Take photographs of the crime scene and surroundings. Be sure to get pictures of the overall setting, the middle range, and close-ups.
4. Determine the best method to search the crime scene. An area of the crime scene will be assigned to each group. Make a sketch of the scene, and indicate on the sketch where each piece of evidence is found.
5. Collect any evidence that needs to be tested. Record what it is, its basic characteristics, and where it was found.
6. Test your various samples, and compare them to the samples provided by your teacher.

Extension Option

As a group, present the evidence you found at the crime scene. Be sure to discuss the various situations that might be responsible for the absence of your teacher.

Data Collection and Analysis

Crime Scene Sketch

Name ______________________________ Date ______________________

20. Summary: Missing Person—Your Teacher!

STUDENT ACTIVITY PAGE

Evidence found	Description	Location collected
1.		
2.		
3.		
4.		
5.		
6.		
7.		
8.		
9.		
10.		

(chart continues on next page)

Name ______________________ Date ______________________

20. Summary: Missing Person—Your Teacher!

STUDENT ACTIVITY PAGE

Evidence found	Description	Location collected
11.		
12.		
13.		
14.		
15.		
16.		
17.		
18.		
19.		
20.		

Name ______________________________ Date ______________________________

20. Summary: Missing Person—Your Teacher!

STUDENT ACTIVITY PAGE

CONCLUDING QUESTIONS

1. Were the stains you collected at the scene blood? If not, what were they?

2. What did the apparent blood spatter indicate about how the blood landed on the surface?

3. Did the signature on the note match the other handwriting samples from your teacher? What does this suggest?

4. Did you identify the pliers that made the tool marks? To whom did they belong?

Name ______________________ Date ______________________

20. Summary: Missing Person—Your Teacher!

STUDENT ACTIVITY PAGE

5. What kind of fibers were left on the lockbox? Is it likely that they were from your teacher's clothing?

____

____

____

____

6. Based on the evidence you collected, was your teacher the victim of foul play? If not, what happened?

____

____

____

____

FOLLOW-UP ACTIVITY

As a class, review the manner in which all of the evidence was collected and tested. What steps would have insured better organization for the collection and treatment of evidence?